7/12 SL
1/17 JL

Wayne
RG133.
Boleyn-Fitzgerald, Miriam
Beginning life
183261514

CONTEMPORARY ISSUES in SCIENCE

BEGINNING LIFE

BEGINNING LIFE

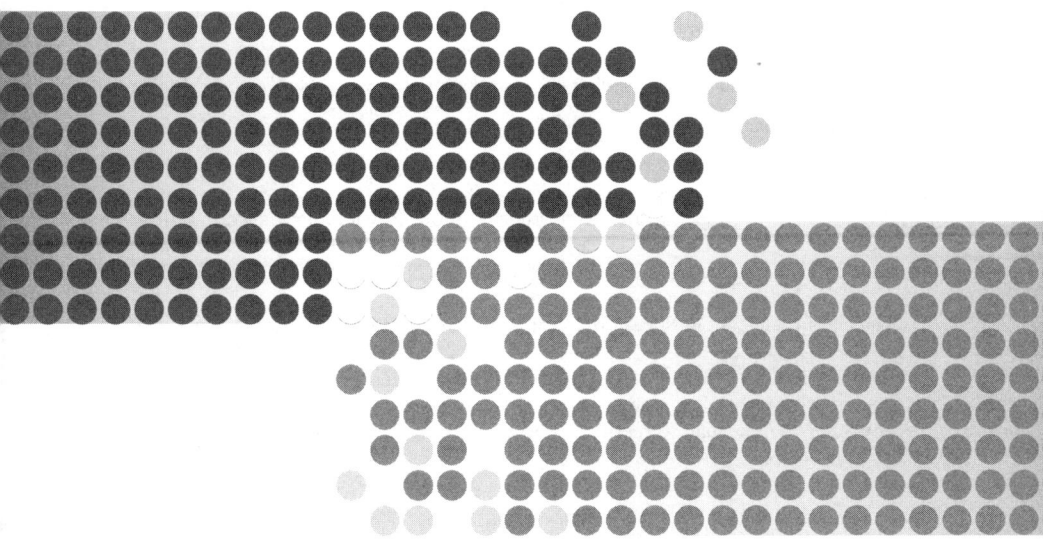

MIRIAM BOLEYN-FITZGERALD

BEGINNING LIFE

Copyright © 2010 by Miriam Boleyn-Fitzgerald

All rights reserved. No part of this book may be reproduced or utilized in any form or by any means, electronic or mechanical, including photocopying, recording, or by any information storage or retrieval systems, without permission in writing from the publisher. For information contact:

Facts On File, Inc.
An imprint of Infobase Publishing
132 West 31st Street
New York NY 10001

Library of Congress Cataloging-in-Publication Data

Boleyn-Fitzgerald, Miriam.
  Beginning life / Miriam Boleyn-Fitzgerald.
    p. cm. — (Contemporary issues in science)
  Includes bibliographical references and index.
  ISBN-13: 978-0-8160-6210-2
  ISBN-10: 0-8160-6210-2
1. Human reproductive technology. 2. Fertilization in vitro, Human. 3. Medical innovations. I. Title.
  RG133.5.F34 2009
  618.1'78—dc22   2008040908

Facts On File books are available at special discounts when purchased in bulk quantities for businesses, associations, institutions, or sales promotions. Please call our Special Sales Department in New York at (212) 967-8800 or (800) 322-8755.

You can find Facts On File on the World Wide Web at http://www.factsonfile.com

Excerpts included herewith have been reprinted by permission of the copyright holders; the author has made every effort to contact copyright holders. The publishers will be glad to rectify, in future editions, any errors or omissions brought to their notice.

Text design by Annie O'Donnell
Illustrations by Melissa Ericksen
Photo research by Tobi Zausner, Ph.D.
Composition by Keith Trego
Cover printed by Art Print, Taylor, Pa.
Book printed and bound by Maple-Vail Book Manufacturing Group, York, Pa.
Date printed: March, 2010
Printed in the United States of America

10 9 8 7 6 5 4 3 2 1

This book is printed on acid-free paper.

CONTENTS

Preface	viii
Acknowledgments	xiii
Introduction	xiv

1. ETHICAL PRINCIPLES IN GENETIC AND REPRODUCTIVE RESEARCH — **1**

Louise Brown: The First Test-Tube Baby	2
The Belmont Report: Ethical Principles in Medical Research	6
The Problem of Scope	7
Patenting Life: The Commercial Ownership of Genes and Organisms	10
In Vitro Fertilization Today	13
Summary	20

2. CONTROVERSIES IN ASSISTED REPRODUCTION — **21**

ART and Multiple Births: The McCaughey Septuplets	21
Maternal and Infant Health Risks of Multiple Births	24
Surrogate Motherhood	28
Selective Reduction: When Surrogates and Intended Parents Disagree	32
Surrogacy Across National Borders	36
Summary	39

3. EUGENICS, GENETIC TESTING, AND DESIGNER BABIES — **40**

Genetic Knowledge, Genetic Control	41

Eugenics as Social Policy: The Carrie Buck Story	46
Contemporary Liberal Eugenics	49
Infant Screening for Treatable Conditions	54
Genetic Discrimination and Privacy	57
Screening Children and Adults for Incurable Diseases	58
Summary	61

4. GENE THERAPY AND ENHANCEMENT — 63
What Is Gene Therapy?	63
The Death of Jesse Gelsinger	67
Status of Gene Therapy Research	69
Technical Limitations and Ethical Concerns	71
The Transhumanist Movement	79
Summary	81

5. STEM CELLS AND THERAPEUTIC CLONING — 82
Types of Stem Cells	83
Turning Adult Cells "Embryonic"	92
Patchwork Policies on Stem Cell Research	96
Celebrities Speak Out on Both Sides of the Debate	98
Embryonic Stem Cells: A Cure for Age-Related Blindness?	101
Summary	102

6. ABORTION AND EMERGENCY CONTRACEPTION — 104
Before Abortion Was Legal: The Sherri Finkbine Story	105
Perspectives on Abortion	106
Human-Animal Hybrids and the Right to Life	108
Emergency Contraception	112
When Abortion Became Legal: *Roe v. Wade*	114

The Partial-Birth Abortion Ban Upheld:
Gonzales v. Carhart   117
Summary   118

7. REPRODUCTIVE CLONING AND ECTOGENESIS   119
Dolly: The First Animal Cloned from an Adult   120
Animal Cloning   123
Cloning Humans?   125
Ectogenesis: Growing Babies Outside the Womb   127
Ectogenesis and the Abortion Debate   129
Summary   131

8. INFANTS   132
Baby Doe and the Treatment of Impaired Infants   133
Extreme Prematurity   137
Baby Messenger and Aggressive Treatment of Premature Newborns   140
Denmark's Required Minimum Gestational Age   142
Summary   144

Chronology   145
Glossary   152
Further Resources   165
Index   183

PREFACE

"Whenever the people are well-informed, they can be trusted with their own government. Whenever things get so far wrong as to attract their notice, they may be relied on to set them to rights."

—Thomas Jefferson

In today's high-speed, high-pressure world, keeping up with the latest scientific and technological discoveries can seem an overwhelming, even impossible task. Each new day brings a fresh batch of information about how the world works; how human bodies and minds work; how human civilization can "work" the world by applying its collective knowledge. Switch on a television news program or the Internet at this very moment—pick up any newspaper or current interest magazine—and stories about health and the environment, worries about national security and violent crime, or advertisements for the latest communication and entertainment gadgets will abound.

Given the nonstop flow of information and commercial pressures, it may seem that a surface understanding of scientific and technological issues is the only realistic goal. The Contemporary Issues in Science set is designed to dispel the myth that a deeper understanding of new findings in science and technology—and therefore considerable power to influence their use—is out of reach of nonspecialists and should be "left to the experts." The set reviews current topics of universal relevance like global warming, conservation, weapons of mass destruction, genetic engineering, medical research ethics, and life extension, and explores—through the lens of real people's stories—how recent discoveries have changed daily life and are likely to alter it in the future.

Preface

Stories featured in the set have received attention in the popular press—often provoking heated controversy at a local, national, and sometimes international level—because beneath the headlines lie sticky questions about how new knowledge should, or should not, be applied, as illustrated by the following examples:

- *Genetic engineering.* The pace of discovery about the human genome and the genomes of other animal and plant species has been breathless since the year 1953, when James Watson and Francis Crick first described the double helix structure of deoxyribonucleic acid (DNA), the chemical substance that acts as a blueprint for building, running, and maintaining all living organisms. In April 2003—a mere 50 years later—sequencing of the human genome was complete. This impressive surge in knowledge about our genes has been accompanied by intense hopes—and intense fears—about newfound technical powers to manipulate the production of life. The tragic death of 18-year-old Jesse Gelsinger in a 1999 gene therapy trial begged obvious questions: Can medical investigators ever obtain truly informed consent from a volunteer when the risks of an experimental procedure are largely unknown? Are the potential benefits of gene therapy worth the unknown public-health risks of altering the human genome using viral vectors? What are the environmental risks of creating transgenic plant and animal species?
- *End-of-life care.* Bold medical innovations like mechanical ventilation, organ transplantation, and tube feeding have saved and improved the lives of millions of patients since the 1950s. A state of profound unconsciousness known as "irreversible coma" first occurred with the ventilator; before its availability, patients without working respiratory systems died from lack of oxygen. Now the bodies of severely brain-damaged and brain-dead people can be maintained indefinitely

with a steady supply of oxygen to their living tissues. Theresa Schiavo's case—and other controversial end-of-life cases—shows how loved ones and medical professionals try to grapple with agonizing questions like: When are medical interventions extending meaningful life, and when are they inappropriately prolonging death? If a patient's wishes cannot be known with certainty, who should decide her fate?

- *Consumer choice.* Using cheap and plentiful energy; selecting personal transportation modes; building and occupying homes; consuming . . . well, just about anything: These options are all realized through technological innovation. Consumer choice is credited for dramatically improving quality of life in North America over the past century, but it has also created a suite of forbidding problems: global climate change, pollution, urban sprawl, and resource depletion. Can modern consumers—especially the rapidly increasing Chinese and Indian "middle-classes"—enjoy the same choices, or the same quality of life, as North Americans of the last half of the 20th century? Will purely technological solutions for problems arise (e.g., will a form of cheap and reliable carbon sequestration be developed to store carbon dioxide, allowing coal to be used to produce cleaner electricity)? Or will technology provide the means for a dramatic change in how people live and work (e.g., will ubiquitous broadband and wireless access lead to the delocalized office—employees always at work, so there is no need to "go to work," no matter where they are)?

- *Water.* With "peak water" (the maximum amount of clean, usable water available globally) predicted to occur sometime in the next 25 years, this vital natural resource is certain to be the source of national and regional conflicts. Water plays an essential role not only in living processes but in industrial-scale heating and cooling and in new alternative energy technologies such as coal gasification, hydrogen production, and biofuels conversion. Water also figures highly in global climate change, acting both as a greenhouse gas and as a dynamic heat reservoir. For humankind's clean water requirements, is technological advancement the problem or is it the solution? Will gigantic energy-efficient desalination plants turn countries with ocean coastlines into the new "wet" OPEC, with "clear gold" (water) replacing "black gold" (petroleum) as the preeminent wealth-generating natural resource? Can technological innovation lessen the terrible toll that floods and droughts take on property and human lives?

- *Privacy.* Today, all bits and pieces of personal information—financial, medical, political, religious, identity-by-association, consumer preference, and lifestyle—are being collected, parsed, amalgamated, mined, and analyzed at a rate, and to an extent, unimaginable a decade ago. An individual's personal information can be collected, shared, exchanged, sold, disseminated, and broadcast without notice given to, or permission received from, the individual—and all perfectly legally. Identity theft is a widespread and growing problem—a phenomenon both created and addressed by modern electronic and software technologies. The use of e-mail to acquire personal financial information under false pretences, known as "phishing," was estimated to have cost U.S. citizens over $2.8 billion in 2006. Can the benefits of instantaneous and remote transactions—financial, consumer-based, social, and educational—ever outweigh the loss of privacy or the risk of being victimized? Who really owns a person's digital identity—the individual, banks, insurance companies, or government agencies?
- *Weapons.* On August 6, 1945, the city of Hiroshima, Japan, was annihilated by an atomic bomb that killed an estimated 70,000 civilians instantly. Radio Tokyo described the extent of the devastation in a broadcast intercepted by Allied forces: "Practically all living

things, human and animal, were literally seared to death." Three days later, a second nuclear bomb was detonated—this time over the southern port city of Nagasaki—killing another 40,000 to 75,000 people. Nuclear weapons have not been used since, but many countries have sought and achieved the technology to deploy them. What is the real threat of nuclear warfare in the early 21st century? What other potentially devastating weapons are being developed today, and how can human civilization avoid its own violent destruction?

Whether readers are students considering a career in a scientific or technical field, science or social studies teachers or librarians, or inquisitive people of any age with personal, professional, or political interests in how new knowledge is applied, the Contemporary Issues in Science set places fresh research findings in the context of real-life stories, clarifying the technical and ethical subtleties behind the headlines and supporting an engaged, informed citizenry.

ACKNOWLEDGMENTS

I would like to extend special thanks to the creative team behind these books: executive editor Frank Darmstadt, for his keen editorial eye; literary agent Jodie Rhodes, for making the match; photo researcher Tobi Zausner, for her talent and tenacity in hunting down pictures; Melissa Ericksen, for her outstanding work on illustrations; Alexandra Lo Re, for her meticulous readings; and Peter Faguy, for his shared vision and passion for the set.

I owe a profound debt of thanks to my parents and to the teachers and professional mentors who encouraged me to clear my own path. It is with boundless gratitude and appreciation that I dedicate the set to my husband, Patrick, and my son, Aidan, for giving me so many good reasons to get out of bed, sit down and write, and love them.

INTRODUCTION

Just before noon on November 19, 1997, Bobbi McCaughey delivered seven infants by cesarean section, setting a world record for the number of babies born alive from one pregnancy. The birth of septuplets is rare, and never before had all seven survived. Dr. Paula Hauser, after she and Dr. Karen Drake delivered the babies, told *Time* magazine, "It just strikes me as a miracle."

While the media focused on their miraculous survival, many medical ethicists and physicians worried about the babies' long-term health. The birth had come nine weeks early, and like most infants born so prematurely, the septuplets could not breathe on their own. Doctors had to place all seven babies on ventilators, and several of them developed serious chronic health problems. For some observers, the McCaughey case was a cautionary tale about the growing trend of multiple births caused by fertility procedures, the resulting flood of tragedies for parents and babies, and the burden placed on society by the high cost of intensive neonatal care.

Before the widespread use of mechanical ventilators, extremely premature babies usually died due to the immaturity of their lungs. The ventilator's arrival in the 1950s offered a new way to sustain babies while their lungs developed, and it created a new need for special neonatal wards and "intensive care" techniques to treat babies living on ventilators. Families and doctors suddenly faced new, often agonizing decisions about whether to withdraw care from babies born severely impaired or underdeveloped.

Beginning Life takes a close look at several bold medical innovations that have created new lives and saved others—assisted reproductive technologies like in vitro fertilization (IVF) and surrogacy, genetic testing and therapy, stem cells and therapeutic cloning, and intensive care techniques for severely premature newborns—

Introduction

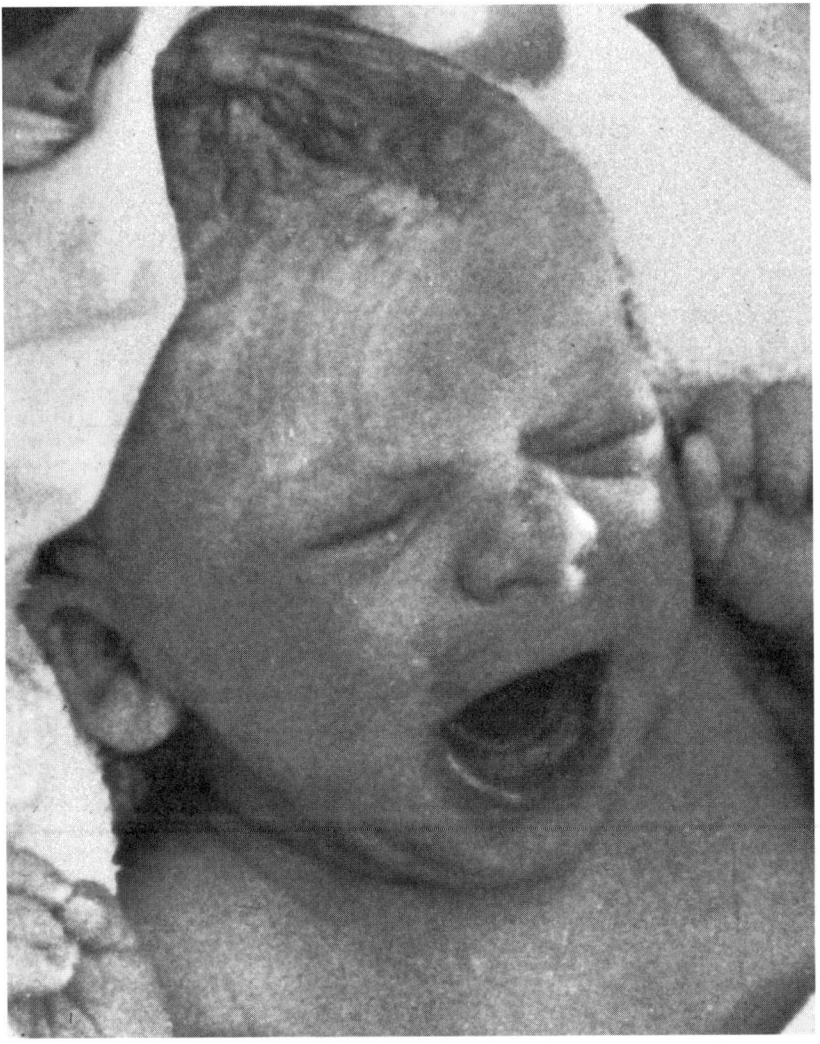

Newborn Louise Brown, the world's first test-tube baby, pictured shortly after she was born at Oldham General Hospital, England, on July 25, 1978 *(AP Images)*

and their exciting and complex implications for hopeful parents, sick patients, loved ones, and health-care practitioners.

Medical innovation requires research, and research depends on patients and healthy volunteers willing to assume risks. *Beginning Life* considers ethical decisions faced by medical researchers on the path to discovery, using patients' stories to highlight key ethical principles and their application to the everyday practice of medicine.

Each story chosen for the book has received attention in the popular press and provoked controversy at a local, national, and sometimes international level. Yet at the heart of each of these stories, underneath the political rhetoric and media hype, are the lives of real patients. Students considering careers in health care, medical research, and medical ethics can immerse themselves in these stories to better grasp the ethical choices involved and the importance of weighing and balancing potential consequences of those choices carefully, on a case-by-case basis.

At the research end of the process, medical ethics tries to balance the risks and potential benefits of experimentation while ensuring that human subjects are truly informed and protected throughout the process—no easy agenda, especially when experimental treatments are extraordinarily new and their risks largely unknown. When medical research succeeds, it creates knowledge, paves the way for future innovation, and leads to longer, healthier lives. At the treatment stage, medical ethics grapples with issues like: When does aggressive treatment of an impaired newborn extend meaningful life, and when does it inappropriately prolong needless suffering? At what point should health-care providers counsel hopeful parents to end expensive, often physically and emotionally taxing fertility treatments? Should fertility specialists allow parents to test embryos for sex and other nondisease-related genetic traits, or should preimplantation genetic diagnosis (PGD) only be used to prevent tragic diseases like Tay-Sachs? Is access to reproductive and genetic technology a right or a privilege, a national or a global issue?

These questions infuse the real people's stories in this volume. They are sticky questions because they reveal people's scientific, ethical, and religious intuitions about when life begins and where a person's individual identity resides—is it in her genes, her personal history, her soul or spirit?—and because different people's intuitions about these questions often come into direct conflict.

Introduction

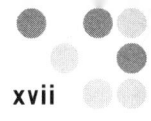

Beginning Life follows the development and interplay of research ethics and of major advances in genetic and reproductive technology from the mid-20th through the early 21st centuries. Chapter 1 tells the story of the first test-tube baby, Louise Brown; traces the development of ethical codes for medical research; and applies ethical principles to recent controversies over IVF. Chapter 2 considers special issues that arise with multiple births and surrogacy; chapter 3 looks at hopes and fears about newfound genetic knowledge and the growing number of tests for particular genetic traits; and chapter 4 treats the tumultuous history and uncertain future of experimental gene therapy. Chapter 5 looks at stunning breakthroughs and recent controversies in stem cell research, and chapter 6 treats the closely related issues of abortion and emergency contraception. Chapter 7 reviews the brief history of animal cloning, perspectives on human cloning, and new developments in artificial womb technology, while chapter 8 brings many of the book's themes together with a look at ethical challenges in the treatment of impaired and severely premature newborns.

Tensions between ideas about what is right and the practical challenges inherent in the real-world practice of medicine will always exist and will continue to generate fruitful debate in the context of fresh medical discoveries. New generations of scientists, ethicists, health-care practitioners, policy makers, and patients will need to turn to individual cases like the ones featured in this book to identify key issues and establish common ground for discussion.

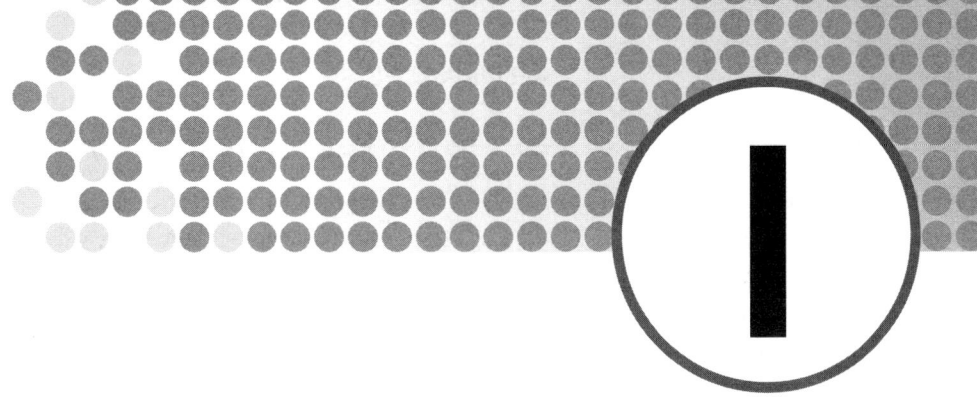

Ethical Principles in Genetic and Reproductive Research

Three decades ago, the first "test-tube" baby was born, much to the fascination of onlookers worldwide. Today more than a million children have been conceived under controlled laboratory conditions. This chapter tells the story of Louise Brown, the first test-tube baby, and reviews early medical and ethical concerns about *in vitro fertilization (IVF)*.

The next part of the chapter provides a framework for evaluating reproductive and genetic research using three key *ethical principles,* and discusses the "problem of scope" (the moral status of *embryos* or *fetuses*)—a flash point of controversy among medical ethicists, researchers, policy makers, religious groups, and advocates for *stem cell research.*

The chapter wraps up with a look at current ethical and health concerns with in vitro fertilization, the most commonly used *assisted reproductive technology (ART)* today.

LOUISE BROWN: THE FIRST TEST-TUBE BABY

At Oldham and District General Hospital in north-central England, the world's first test-tube baby was born to proud parents John and Lesley Brown on the evening of July 25, 1978. Her name was Louise, and to most who followed the event, her arrival seemed nothing short of a miracle.

Louise's parents were ecstatic. "It was like a dream," her father told a reporter for London's *Daily Mail*. Her mother said, "She's so small, so beautiful, so perfect." The *Mail* had purchased the rights to Louise's story and pictures for a rumored $570,000, but other newspaper and television reporters swamped the hospital, hoping to personalize their takes on the family's story.

A now-famous video of Louise taking her first breaths was televised all over the world. Many of the stories had miraculous—even religious—undertones. *Newsweek* likened her birth to "a first coming," and the *Daily Express* headline went so far as to suggest that she held "The Whole World in Her Hands."

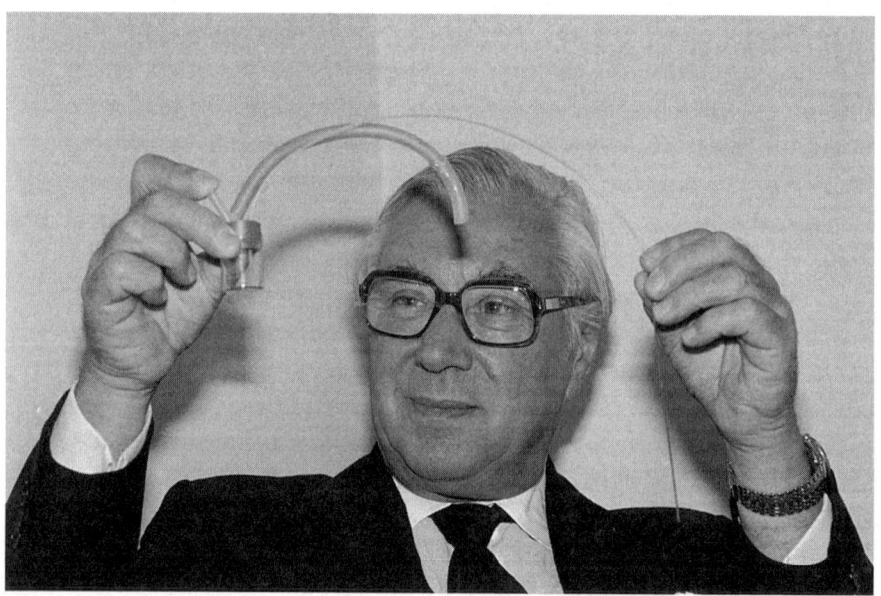

Dr. Patrick Steptoe, one of the pioneers of in vitro fertilization, holding a device like the one used to remove a mature egg from Mrs. Lesley Brown's ovary for fertilization in the laboratory *(Bettmann/CORBIS)*

Ethical Principles in Genetic and Reproductive Research

The Browns had tried to conceive for almost a decade before attempting to adopt a baby. Two years later, they were still on the waiting list. A nurse referred them to Robert Edwards (an embryologist) and Patrick Steptoe (a gynecologist) who had been working together for years to perfect the surgical and biochemical techniques necessary for in vitro fertilization.

In vitro literally means "in glass"—performed inside a test tube or other laboratory apparatus. After a woman undergoes a course of hormonal treatments, a ripe *ovum* (egg) is removed from her ovary and placed in a solution of nutrients and sperm to be fertilized. Once the ovum has begun to divide into multiple cells, it is implanted in the womb.

"The last time I saw the baby," Dr. Edwards told the *Mail,* "it was just eight cells in a test tube. It was beautiful then, and it's still beautiful now." Dr. Steptoe, who delivered Louise, told a press conference, "She came out crying her head off, a beautiful normal baby."

Despite many doom-and-gloom predictions by opponents of the procedure, Louise was born apparently healthy. When she was older, she did report feeling socially isolated at times, and frequently had to explain to other children that she was not actually born in a laboratory.

Within two years, the first IVF program in the United States was launched at Eastern Virginia Medical School. Many other programs followed, but some localities chose not to publicize them. Other plans for clinics were cancelled due to opposition from religious and right-to-life groups. Many people equated the procedure with *abortion,* since only the embryos that appeared to be developing normally were transplanted from the laboratory environment to the womb. Defenders of the procedure argued that it only mimicked what would happen naturally to embryos (i.e., miscarriage) were they not dividing normally. (There are additional ethical concerns today with the now-common practice of producing "spare" embryos for implantation; see the section later in this chapter, "In Vitro Fertilization Today.")

Though Louise Brown was born healthy, IVF procedures remained controversial both on medical and ethical grounds. News stories at the time were flush with concern about the

Louise Brown, with her parents, Lesley and John Brown, on the *Phil Donahue Show* in September 1979 *(Bettmann/CORBIS)*

potential harm to babies. Cautioned *Newsweek,* "What if the learning process leads to mistakes, in the form of test-tube babies born with mental or physical defects?" The news magazine pointed to well-known lawyers, one of whom envisioned "malpractice suits by such children against either their parents or the physicians who created them."

Though such nightmare scenarios—to anyone's knowledge—never came to pass, some experts still argue that the health impacts of IVF procedures for children and mothers have never been properly evaluated. Other ethical reservations raised at the time of Louise Brown's birth remain unresolved, and they are at the heart of some of the most passionate objections to assisted reproduction and medical research with embryos. Children have not yet been grown outside the womb, nor has a generation of super babies been genetically engineered, but these possibilities remain real and controversial.

Father William Smith of New York's Catholic archdiocese said in 1978 of Louise Brown's exceptional birth, "I fear that we may be slipping away from doctoring the patient to doctoring the race." The Roman Catholic position on IVF remains the same today as in 1978, holding that the procedure is morally wrong, since it often results in the destruction of embryos, and since it replaces the "natural" conjugal union between husband and wife. The church officially places IVF in the same category as *artificial insemination,* which Pope Pius XII condemned in 1951 because it "converts the domestic hearth, the sanctity of family into nothing more than a biological laboratory."

Not every religious commentator at the time came out against the procedure. Arthur Dyck, a United Church of Christ layman and professor of ethics at Harvard University, told *U.S. News and World Report,* "No one says we should meekly submit to natural disasters such as hurricanes. Nature sets limits, but it's our task to improve on nature and try to perfect the process because we value the life that God has given us."

Other medical scenarios that at the time were characterized as the stuff of science fiction have since become quite routine. Here is an excerpt from *Newsweek* just after Louise Brown's birth: "[T]here are widespread misgivings over the next possible steps: surrogate mothers who might rent out their wombs . . ." And from *U.S. News & World Report:* "This conjures up the nightmarish scenario of a generation of test-tube babies searching out egg and sperm banks for their origins—and updating an old line to: 'It's a wise child who knows its own mother.'" *Surrogate motherhood,* of course, has been routine practice for years (see chapter 2), as has the use of egg and sperm donors for IVF.

Again, from *Newsweek:* "A group of scientists . . . reported last week that they had identified, for the first time, a single gene among the millions in one human cell. Their technique derives from the controversial tinkering with heredity known as recombinant DNA technology. Their finding promises the possibility of detecting genetic diseases in fetuses still in the womb. But the same recombinant technology might eventually be used to alter the genes of human fetuses just fertilized in the test tube."

The screening of embryos for genetic diseases—even for sex—has become quite common in recent years (see chapter 3), and though *genetic engineering* of embryos has not yet occurred, public attitudes toward the prospects have softened. A recent poll conducted by Virginia Commonwealth University found that 41 percent of Americans would be open to using genetic engineering to reduce their children's risk of serious disease. (The issues of genetic therapy and enhancement will be taken up in chapter 4.)

Arthur Caplan, director of the Center for Bioethics at the University of Pennsylvania, told PBS's *American Experience* in 2006, "The day Louise Brown got made was the day that a core aspect of human life, reproduction, moved from a mystery to a technology, moved from something that we were in awe of to something that we manipulate . . . [T]here's nothing more basic you're going to change before or since in the history of humanity."

THE BELMONT REPORT: ETHICAL PRINCIPLES IN MEDICAL RESEARCH

In 1979, the National Commission for the Protection of Human Subjects of Biomedical and Behavioral Research published its "Belmont Report," which laid out three basic ethical principles:

1. *Respect for Persons.* The commission recognized the importance of preserving personal *autonomy,* or the ability of a person to make independent choices, primarily through guidelines for informed consent. "Respect for persons," the commission wrote, "requires that subjects, to the degree that they are capable, be given the opportunity to choose what shall or shall not happen to them. This opportunity is provided when adequate standards for informed consent are satisfied." Though exact requirements for informed consent are debatable and difficult to standardize, the commission broke the consent process down into three basic elements: information, comprehension, and voluntariness.
2. *Beneficence.* Fulfilling this principle means both minimizing possible risks and maximizing possible benefits to research subjects. The commission recognized the challenge in "making precise judgments" about risks and benefits when those risks and benefits cannot

always be known but stressed that "the idea of systematic, nonarbitrary analysis of risks and benefits should be emulated insofar as possible." In other words, researchers should do their absolute best to ensure that their subjects' interests are served.
3. *Justice.* The equal or fair distribution of the burdens and benefits of research was also at issue. The commission was troubled that in the past, "some classes" of people—for example, minorities, prisoners, and the economically disadvantaged—had been "systematically selected" for research "simply because of their easy availability, their compromised position, or their manipulability," and it recommended that publicly funded research "not provide advantages only to those who can afford them" and "not unduly involve persons from groups unlikely to be among the beneficiaries of subsequent applications of the research."

These principles form a framework for consideration of potential harms and benefits in *genetic testing* and experimental therapy with children and adults, as well as risks to parents or children posed by new assisted reproductive technologies. Many of the ethical debates considered in this volume hinge on the application of these three principles. Can a person who is considering genetic testing foresee the potential psychological consequences of learning that he has a deadly gene, and therefore give truly informed consent? Can the risks of IVF be adequately described to a woman weighing and balancing the pros and cons of the procedure, when no systematic study of its effects exists? And can surrogate mothers be protected from exploitation when money changes hands, especially across national borders? These are tough questions with serious implications for the health and well-being of patients, parents, children, and future children.

THE PROBLEM OF SCOPE
The three major principles outlined above should guide any genetic or reproductive research involving babies, children, or

adults, but when applying these principles in medical research, should human embryos or fetuses be considered to have the same *moral status* as humans, no moral status at all, or something in between? There is no consensus on this issue within the broader culture, nor is there agreement among medical ethicists and researchers.

This cultural stalemate is evident in the wide range of opinions expressed by members of the President's Council on Bioethics in their report, "Human Cloning and Human Dignity: An Ethical Inquiry." President George W. Bush established the council to consider the consequences of stem cell research, and while its members unanimously concluded that *cloning-to-produce-children* (alternatively—and sometimes controversially—termed *reproductive cloning*) is unethical, the council was split on the issue of

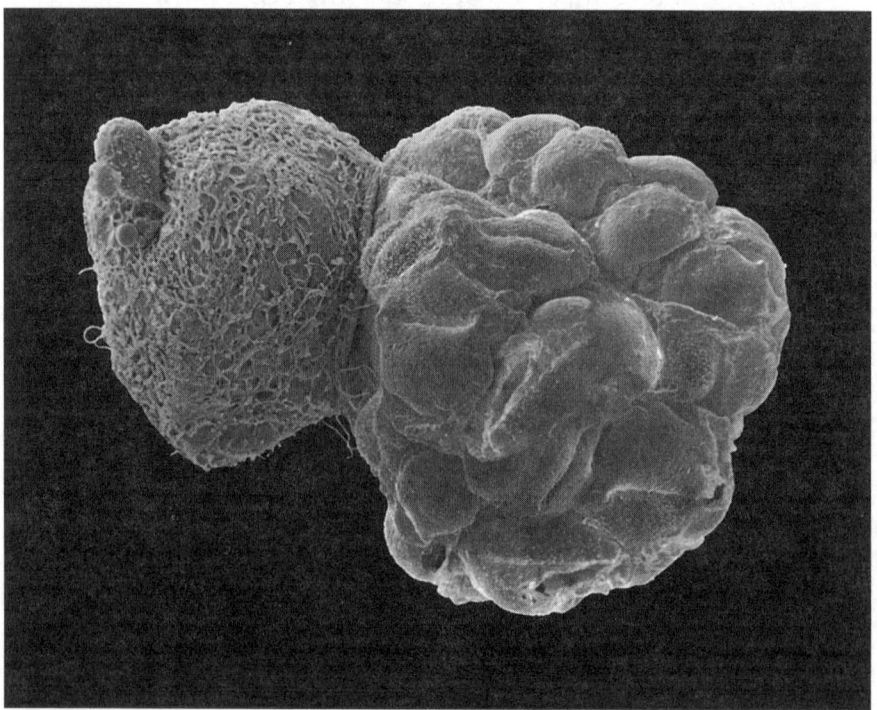

A developing human embryo at the blastocyst stage, five days after fertilization, seen here "hatching" from the protein shell that surrounds the unfertilized egg *(Dr. Yorgos Nikas/Photo Researchers, Inc.)*

Ethical Principles in Genetic and Reproductive Research

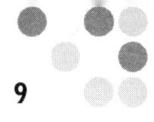

whether *cloning-for-biomedical-research* (or *therapeutic cloning*) should be allowed.

Some council members supported cloning-for-biomedical research on the grounds that "it may offer uniquely useful ways of investigating and possibly treating many chronic debilitating diseases and disabilities, providing aid and relief to millions." They expressed two distinct positions on the moral status of embryos:

1. *Early-stage embryos have intermediate moral status.* "While we take seriously concerns about the treatment of nascent [early] human life," wrote these council members, "we believe there are sound moral reasons for not regarding the embryo in its earliest stages as the moral equivalent of a human person. We believe the embryo has a developing and intermediate moral worth that commands our special respect, but that it is morally permissible to use early-stage cloned human embryos in important research under strict regulation." Specifically, these council members recommended limiting research to the first 14 days of development, before any organ differentiation occurs.
2. *Early-stage embryos have no moral status.* These council members concluded that early-stage cloned embryos "should be treated essentially like all other human cells" and that "the moral issues involved in this research are no different from those that accompany any biomedical research," e.g., the need to obtain informed consent from the donors of eggs and *somatic cells* used to clone embryos for research.

A third position on the moral status of embryos was expressed by council members who opposed cloning-for-biomedical-research. "We find it disquieting," they wrote, "even somewhat ignoble, to treat what are in fact seeds of the next generation as raw material for satisfying the needs of our own." They took the position that:

(continues on page 12)

PATENTING LIFE:
The Commercial Ownership of Genes and Organisms

The problem of scope hinges on the question, "When does life begin?" and beliefs on this issue vary drastically, but the idea of patenting (owning the commercial rights to) a human at any stage of development—including embryonic and fetal stages—is generally considered offensive. Until the last few decades, U.S. legal doctrine has prohibited patenting any products of nature, but in 1980 the U.S. Supreme Court broke with this tradition by ruling that a genetically engineered bacterium capable of breaking down crude oil was patentable. Chief Justice Warren Burger wrote that "the fact that micro-organisms are alive is without legal significance" and that patent law covers "anything under the Sun that is made by man." Because the organism had been genetically modified, argued the 5-to-4 majority, it no longer counted as a product of nature.

This trend continued in 1988, when the U.S. Patent Office granted patent number 4,736,866 to the President and Fellows of Harvard College for OncoMouse—a genetically modified mouse that was extraordinarily susceptible to cancer. OncoMouse (from *onco,* Greek for "tumor") had been created by researchers at Harvard by introducing a gene that encourages tumor growth (*oncogene*) into a fertilized mouse embryo. The Patent Office's decision was in keeping with its new policy of taking "non-naturally occurring non-human multicellular living organisms, including animals, to be patentable subject matter."

The controversial practice of patenting living organisms expanded throughout the 1990s to include vast numbers of genes—the chemical blueprints for all living organisms. Biotech companies, universities, and research institutions now own approximately one-fifth of the genes that build and maintain the human body, including genes that can cause obesity, cardiovascular disease, asthma, and breast cancer. Though naturally occurring DNA sequences are certainly considered "products of nature," patents are issued for the chemical sequences discovered in the technical process of decoding DNA. The holder of exclusive patent rights can charge other laboratories licensing fees to use the information in clinical testing.

Michael Crichton, doctor and author of such well-known medical fiction as *Jurassic Park* and *ER,* noted in the *New York Times* in February 2007 that gene patents "slow the pace of medical advance on deadly diseases. And they raise costs exorbitantly: a test for breast cancer that could be done for $1,000 now costs $3,000. Why? Because the holder of the gene patent can charge whatever he wants, and does. . . . He owns the gene. Nobody else can test for it. In fact, you can't even donate your own breast cancer gene to another scientist without permission. The gene may exist in your body, but it's now private property."

Peter Shorett, director of programs for the Council for Responsible Genetics, calls patents on genes and organisms a "'toll booth' through which future scientists must pass" and says that the higher the cost of obtaining model organisms like OncoMouse, "the more biomedical innovations will be impeded, as researchers in the early stages of their work may choose to look elsewhere, not willing to pay steep up-front costs or abide by unyielding restrictions." The patent process also hinders a primary mission of universities, Shorett argues—the free and open exchange of knowledge. "Secrecy and under-communication become the norm as faculty members withhold data from the scientific community to protect proprietary interests."

Claims that gene patents hinder medical progress are bolstered by a 2002 study by Jon Merz and colleagues at the Center for Bioethics at the University of Pennsylvania, published in the journal *Nature,* which looked at the development of genetic testing for a fairly common and treatable inherited disease known as *haemochromatosis.* Upwards of 80 to 85 percent of haemochromatosis cases are caused by the two most common mutations of a single gene, the HFE gene. Mercator

(continues)

> *(continued)*
> Genetics was granted U.S. patents for the HFE genetic test in 1998, thus allowing the company to exclude others from testing for the two mutations, and Merz and his colleagues found that although "many U.S. laboratories began genetic testing for haemochromatosis before the patents were awarded... 30% of those in our survey reported discontinuing or not developing genetic testing in the light of the exclusive license granted on the patents covering clinical-testing services."
>
> In February 2007, Representative Xavier Becerra, a Democrat from California, and Representative Dave Weldon, a Republican from Florida, introduced the Genomic Research and Accessibility Act in the U.S. House of Representatives, a bill that would prohibit the patenting of genetic material. The act stalled later that winter in the House Subcommittee on Courts, the Internet, and Intellectual Property.

(continued from page 9)
3. *Embryos at any stage of development are human life.* These council members asserted a "continuous history of human individuals from the embryonic to fetal to infant stages of existence" and that if "from one perspective the view that the embryo seems to amount to little may invite a weakening of our respect, from another perspective its seeming insignificance should awaken in us a sense of shared humanity and a special obligation to protect it."

The majority of council members did not recommend banning cloning-for-biomedical-research outright, but instead called for a four-year *moratorium* to allow for public debate and review of research involving embryos. The rest of the council recommended that cloning-for-biomedical research proceed under strict government regulations.

Though a moratorium never was imposed, President George W. Bush vetoed the Stem Cell Research Enhancement Act (H.R. 810) in July of 2006, a bill that would have reversed the federal law making it illegal for federal money to be used in research where *stem cells* are obtained from the destruction of an embryo. Many state governments and universities have worked around these federal restrictions to provide their own funding for stem cell research. (See chapter 5 for a more detailed account of state laws promoting or prohibiting such research.)

The problem of scope will be revisited in the chapters to come, as it is central to any discussion involving abortion, research with embryos, or assisted reproductive technologies like IVF.

IN VITRO FERTILIZATION TODAY

In the early years of in vitro fertilization (IVF) treatment, fears were widespread that test-tube babies would be born monstrously deformed, or at the other end of the spectrum, that parents would misuse the new technology to create super babies. The potential pitfalls of IVF became fair game for late-night talk show humor; Johnny Carson joked that on Father's Day, test-tube babies would be required to send cards to the DuPont Corporation.

Public attitudes have shifted dramatically over the years as the procedure has become more common. The Centers for Disease Control (CDC) estimate that more than one percent of U.S. births are thanks to some kind of assisted reproductive technology (ART)—of which IVF represents more than 99 percent.

"That's sort of the pattern that it takes," said Robin Marantz Henig, author of the book *Pandora's Baby*, interviewed for a 2006 episode of the PBS series *American Experience*. "[A]t first it seems like it's abhorrent and it's something that we absolutely shouldn't do. And then for a while it seems kind of miraculous. And then after a while, the technology just becomes part of the fabric of daily life."

The number of babies born as a result of ART procedures has risen steadily over the almost three decades that IVF has been available in this country. In 2006, a total of 138,198 ART cycles were performed at 426 reporting clinics in the United States,

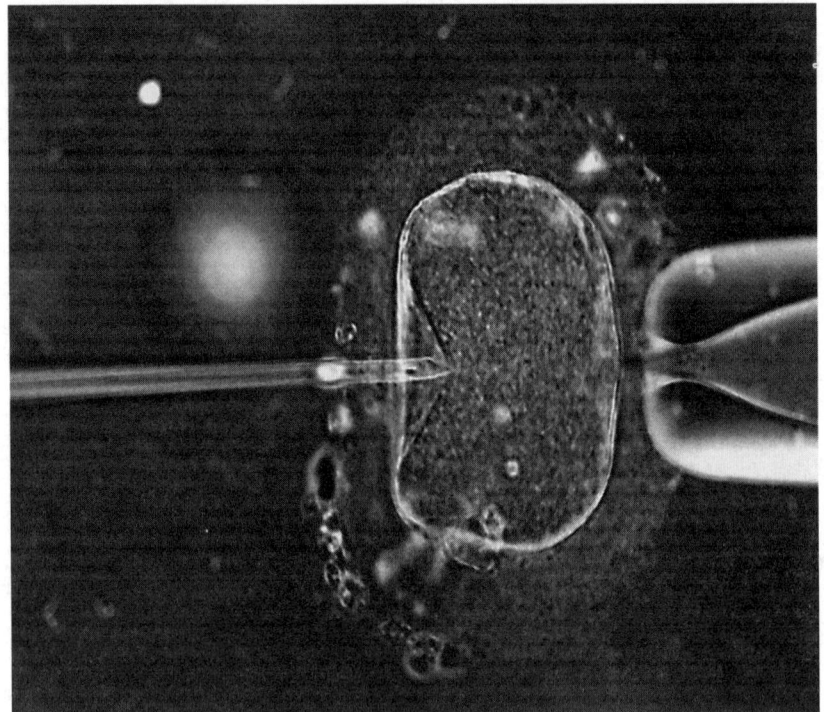

A light micrograph of a type of in vitro fertilization known as intracytoplasmic sperm injection, or ICSI, in which a single sperm is injected into the egg with a microneedle *(Zephyr/Photo Researchers, Inc.)*

resulting in 41,343 *live births* and 54,656 infants. The last two numbers differ markedly because many births from IVF are *multiple births;* it is common practice to implant multiple embryos to increase chances of successful pregnancy.

The number of ART cycles performed in the United States has more than doubled since 1996, when 64,681 cycles were

---

(opposite page) A graph showing the steady increase in ART procedures from 1996 to 2006. One line represents the number of ART cycles performed; one line represents the number of live-birth deliveries, including multiple births; and one line represents the total number of infants born using ART. *(Source: CDC, 2006 Assisted Reproductive Technology Report)*

reported, and the number of live births in 2006 was more than two-and-a-half times higher than in 1996. As IVF techniques have been refined over the years, successful pregnancies rates due to IVF have increased (see figure).

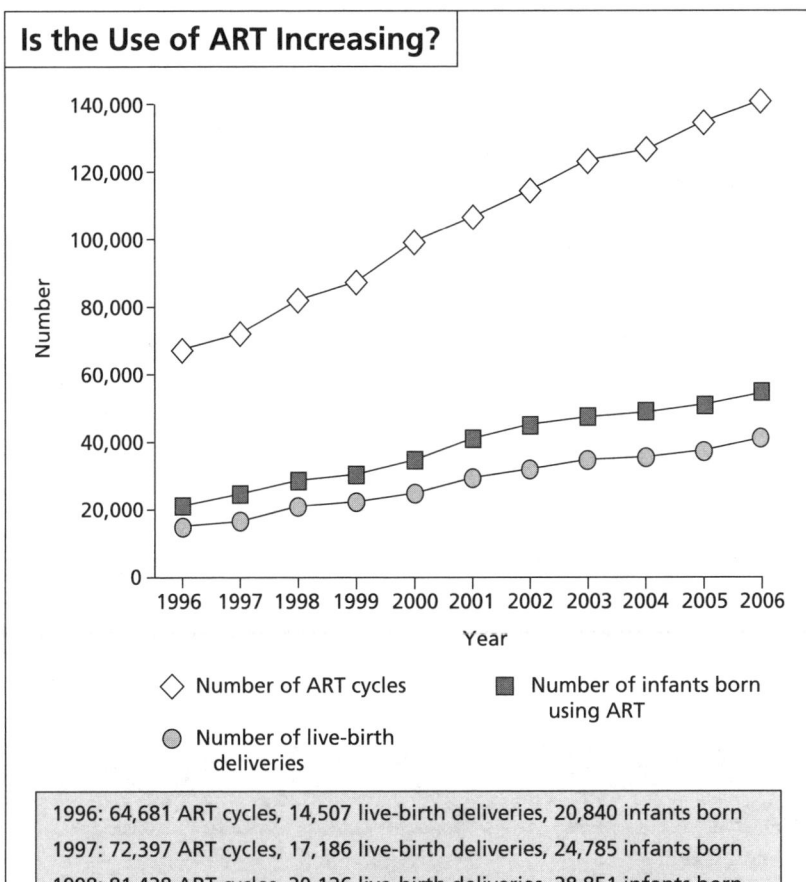

Is the Use of ART Increasing?

◇ Number of ART cycles   ■ Number of infants born using ART
● Number of live-birth deliveries

1996: 64,681 ART cycles, 14,507 live-birth deliveries, 20,840 infants born
1997: 72,397 ART cycles, 17,186 live-birth deliveries, 24,785 infants born
1998: 81,438 ART cycles, 20,126 live-birth deliveries, 28,851 infants born
1999: 87,636 ART cycles, 21,746 live-birth deliveries, 30,629 infants born
2000: 99,629 ART cycles, 25,228 live-birth deliveries, 35,025 infants born
2001: 107,587 ART cycles, 29,344 live-birth deliveries, 40,687 infants born
2002: 115,392 ART cycles, 33,141 live-birth deliveries, 45,751 infants born
2003: 122,872 ART cycles, 35,785 live-birth deliveries, 48,756 infants born
2004: 127,977 ART cycles, 36,760 live-birth deliveries, 49,458 infants born
2005: 134,260 ART cycles, 38,910 live-birth deliveries, 52,041 infants born
2006: 138,198 ART cycles, 41,343 live-birth deliveries, 54,656 infants born

© Infobase Publishing

Despite media hype about much older women using IVF to conceive children, women age 35 or younger received 39 percent of all ART cycles performed in 2006. The average age of women using ART procedures in 2006 was 36 (see the following table). As maternal age increases, so does the use of donated eggs. Four percent of women under 35 used eggs from a donor in 2006, while 21 percent of women ages 41 to 42, and more than half (55 percent) of women over 42, used donated eggs. A full 12 percent of all ART cycles used donated eggs or embryos in 2006, up from 8 percent in 1996.

Despite widespread acceptance of IVF and other assisted reproductive technologies in the broader culture, they remain highly controversial in certain religious circles. Kelly Romenesko, a French teacher from Wisconsin, made national headlines in the fall of 2006 when she was fired from two Roman Catholic schools for conceiving twins through IVF. She and her husband were quick to say that they had used their own eggs and sperm and had not destroyed embryos in the process, but the schools stood by their decision.

Objections from religious leaders notwithstanding, destruction of embryos for IVF and other assisted reproductive processes has never received criticism as severe as the destruction of embryos through abortion. Arthur Caplan noted the inconsistency to PBS's *American Experience:* "I've always been

USE OF ASSISTED REPRODUCTIVE TECHNOLOGY BY MATERNAL AGE

Age Group	Percent of All ART Cycles
<35	39
35–37	23
38–40	19
41–42	10
>42	10

Source: CDC, *2006 Assisted Reproductive Technology (ART) Report* (December 2008)

fascinated," said Caplan, "about the fact that no one, to my knowledge, has ever demonstrated, picketed, chained themselves to the doorway of an in vitro fertilization clinic. And it's not because there have not been condemnations from important religious leaders about the immorality of test tube baby technology ... Test tube baby technology is seen by almost every American as pro-life technology."

As Caplan pointed out, even President George W. Bush—who is morally opposed to the destruction of embryos and who blocked federal funding for stem cell research using so-called "leftover" embryos—has never made a move to stop the destruction of embryos in infertility clinics. "He can't," said Caplan. "He would become persona non grata all over the country if he stepped in and said, 'I'm sorry, infertile people, you can't destroy embryos in the process of trying to have children.' That's just not a politically viable point of view."

400,000 Spare Embryos and Counting

There are vast numbers of frozen embryos stored at hundreds of fertility clinics around the country—some estimates place the number at 400,000 and rising—and what to do with these embryos remains one of the most contentious issues with assisted reproduction today. The choice facing many couples—whether to discard the unused embryos or to donate them to medical science—can be emotionally and morally difficult. Many couples report not anticipating the difficulty of this decision before deciding to move forward with assisted reproduction.

One set of parents who conceived through IVF told NPR in May 2006, "When you walk into a fertility clinic, you have a tendency to focus on 'baby,' and you don't really think about repercussions of the process...." Another couple reported a similar experience: "I think we were so focused on just getting pregnant that we didn't focus on extra embryos at all."

The two couples ultimately made radically different decisions about what to do with the embryos. The first couple donated their nine embryos to another infertile couple through a so-called embryo adoption program (requiring that the adoptive couple sign a document stating that they would

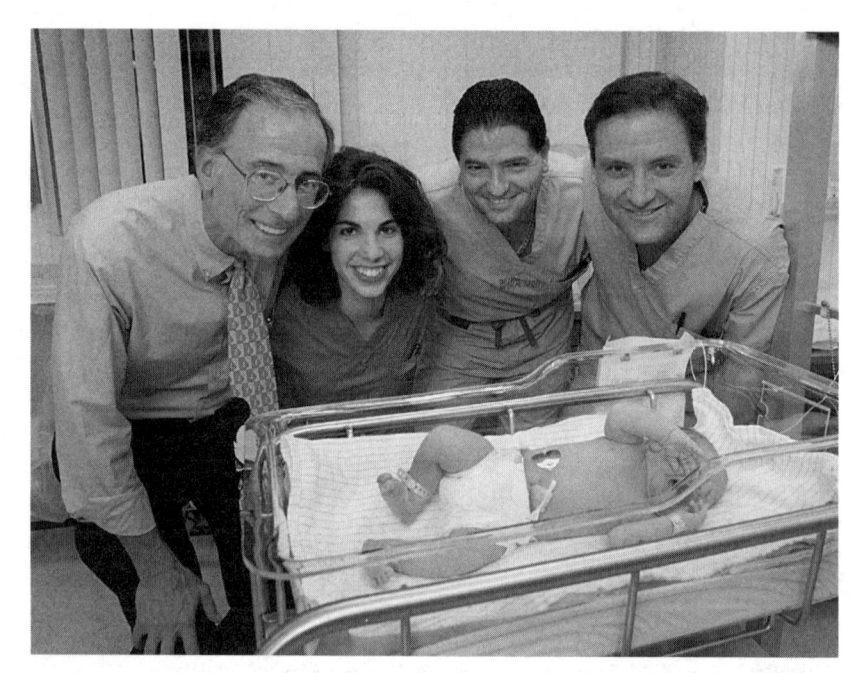

A team of physicians at the Encino-Tarzana Regional Medical Center in Los Angeles stands over the world's "oldest newborn"—a baby boy born in 1998 after developing from an embryo that was kept on ice for seven years. *(AP Images)*

implant all nine embryos), while the second couple donated their three embryos to a research program working to perfect techniques for *pre-implantation genetic diagnosis (PGD)*. Though they chose different fates for their embryos, it was important to both couples that the embryos serve a purpose rather than simply being destroyed.

The High Price of Eggs

The sale of donated eggs is another source of passionate controversy, with compensation for donated eggs soaring in recent years. The price of eggs has increased due largely to demand outpacing supply, but in some cases, exorbitant sums are being paid to donors who possess traits that are seen as desirable. A report published in the May 2007 issue of *Fertility and Sterility*, "What Is Happening to the Price of Eggs?" reported that the average price paid for eggs was $4,217. One donation center reported paying as much as $15,000.

There have been adds in the *Chicago Maroon* (University of Chicago's school paper) offering $35,000 for a Chicago egg, and in the *Harvard Crimson* offering $50,000 for an egg from a Harvard woman. Of obvious concern is whether women in need of funds will feel coerced into donating their eggs.

"The real issue is whether the money can cloud someone's judgment," ethicist Josephine Johnston with the Hastings Center told the *New York Times* in May 2007. "We hear about egg donors being paid enormous amounts of money, $50,000 or $60,000. How much is that person actually giving informed consent about the medical procedure and really listening and thinking as it's being described and its risks are explained?"

Similar concerns have been expressed about affluent couples paying surrogate mothers in low-income countries to carry their children; this issue will be discussed in chapter 2.

Is In Vitro Fertilization Safe?

Another serious concern with IVF and other assisted reproductive technologies is whether the medical risks of in vitro fertilization—either to babies or to mothers—are truly known. Higher rates of multiple births with IVF mean that more babies are born prematurely with lower birth weights, and these factors are associated with higher rates of prenatal, neonatal (newborn), and infant mortality and impairment. (See chapter 2 for a discussion of the McCaughey septuplets, the most famous American example of multiple births from IVF.)

There is some anecdotal evidence to suggest that children born from IVF have higher rates of some *congenital* abnormalities (conditions present at birth) such as *spina bifida* and serious heart defects, but there have been few long-term follow-up studies. Australia is the only country that has kept follow-up data on the health of children born from IVF since the technique was first used, and that data does indicate higher rates of birth defects. One small U.S. study, however, found no evidence of higher rates of birth abnormalities with children born from IVF.

"I think the field ducked the issue [of safety] in part because they didn't do long-term follow-up studies," Arthur Caplan told PBS's *American Experience*. "They basically said, 'These babies look fine.' I remember Steptoe and Edwards sort of displaying Louise Brown around and saying, 'Look, all her parts are here. She's happy . . . We've got a technique that's going to work.' But that didn't prove that there wasn't premature death or unusual amounts of morbidity in these kids."

Robert George, professor of law at Princeton, agrees. "[I]t's critically important that we do the studies that will enable us to know whether in fact, over the long term, there are higher rates of disease, morbidity, among children conceived in IVF. It's very important for the future that potential parents who are contemplating the use of in vitro fertilization know what the potential health risks, if any, are for the children who they will conceive."

SUMMARY

Louise Brown's arrival ushered in an era of hope for millions of infertile couples all over the world. It also set the stage for developments in reproductive and genetic science that seemed far-fetched at the time—*genetic screening* of embryos, *gene therapy*, embryonic stem cell research, egg and sperm donors for hire, surrogate motherhood—many of the new technological practices considered in this book.

Ethical principles used to evaluate biomedical research (respect for persons, beneficence, and justice) remain constant, but unknown risks of many of these new technologies—not only to research subjects, but in some cases to their descendants and contemporaries—mean new practical challenges to the imperatives of informed consent and nonharming.

Moreover, there is no clear cultural consensus on whether embryos or early fetuses should be considered to have equal moral status to fully formed humans, no moral status at all, or some degree of intermediate moral status. This hot-button issue will resurface time and again in the chapters to follow.

Controversies in Assisted Reproduction

Chapter 1 gave an overview of ethical issues with assisted reproduction in the context of the most commonly used ART procedure, in vitro fertilization (IVF). This chapter takes a closer look at how the key ethical principles outlined in that chapter are relevant to two contemporary debates over assisted reproductive techniques: the problem of health risks to infants and mothers due to multiple births and the problem of how best to protect the health and autonomy of women considering surrogate motherhood.

Some opponents of ART procedures and surrogacy—many Catholic theologians, in particular—argue against them on the grounds that they produce human life outside so-called natural reproductive processes. Most mainstream medical debates, however, tend to center on maternal and infant health and how best to protect it.

ART AND MULTIPLE BIRTHS: THE MCCAUGHEY SEPTUPLETS

Bobbi McCaughey (pronounced McCoy) gave birth to the miracle babies of the century—christened the "magnificent seven"

The McCaughey septuplets on their first birthday *(B.KRAFT© 1998/CORBIS SYGMA)*

by many reporters. She delivered the infants by *cesarean section* just before noon on November 19, 1997, setting a world record for the number of babies born alive from one pregnancy. Before they knew it, Bobbi, Kenny, their daughter Mikayla, and her seven new siblings were bathed in the media spotlight, landing on the covers of magazines like *Time* and *Newsweek* and featured on morning news shows. The seven babies became instant media darlings, and their parents were celebrated for making it through a very challenging pregnancy. *People* magazine, for instance, said, "defying the medical odds, Bobbi and Kenny McCaughey clung to their faith and were rewarded—times seven."

The story started when Bobbi and Kenny decided that they wanted a sibling for their sixteen-month old daughter (with whom Bobbi had become pregnant after receiving a fertility drug). Seeking help from a fertility clinic once again, Bobbi received a shot of a strong fertility drug that caused her ovaries to release several eggs on the first try, resulting in a multiple pregnancy. Six weeks later, the first *ultrasound* revealed that

Bobbi was carrying seven fetuses, and the doctor recommended that some of the fetuses be aborted—a procedure known as *selective reduction*—to increase the chances that the remaining fetuses would be born healthy. The McCaugheys rejected this option on religious grounds, and chose to try to carry and deliver all seven.

The next several months were extremely difficult for Bobbi. To reduce the risk of miscarriage, she was placed on bed rest starting in the 19th week of her pregnancy, and she was eventually confined to the hospital. The babies were born at 31 weeks (*full term* is 40) and were all severely underweight, with birth weights ranging from two pounds, five ounces to three pounds, four ounces. All were immediately placed on *ventilator* support. Thanks to a dedicated team of more than 40 specialists, all of the babies survived, albeit some with serious chronic health issues. As of their fourth birthday, one child suffered from *seizures;* two others had forms of *cerebral palsy* (a brain condition that effects muscle coordination and movement); and two still required surgically implanted *feeding tubes.*

The American media focused largely on the miraculous multiple birth and on the parents' honest, hardworking characters and strong faith, as well as on the generosity of others who showered the family with gifts. (The governor of Iowa, for instance, built them a large house to replace the two-bedroom home they had obviously outgrown.) Medical ethicists and physicians, meanwhile, tried to voice concerns about what they saw as the medical mismanagement that led to the multiple pregnancy in the first place—a pregnancy that could easily have resulted in tragedy for the McCaugheys and cost the American health-care system an estimated $1.5 million in premature *neonatal* care.

More than one out of every three (35 percent) of all ART births are multiple, and 4.3 percent are triplets or higher-order multiples. An unknown number of multiple births result from the use of fertility drugs, or *controlled ovarian hyperstimulation (COH)*. Although COH is often included in the broader category of ART, it is not generally reported to the CDC as are more involved laboratory procedures like IVF.

MATERNAL AND INFANT HEALTH RISKS OF MULTIPLE BIRTHS

The number of multiple births in the United States quadrupled between the years 1972 and 1998, largely due to increased use of ART to produce pregnancy. During the 10-year period before the septuplets were born, the number of women administered fertility drugs nearly tripled. Women carrying multiple fetuses have a 3.6 times higher risk of dying from complications than women carrying so-called singleton pregnancies, largely due to fatal blood clots (*embolisms*), dangerously high blood pressure (*pre-eclampsia*), severe bleeding (*hemorrhage*), and infections. Women carrying triplets and higher-order multiples are also at

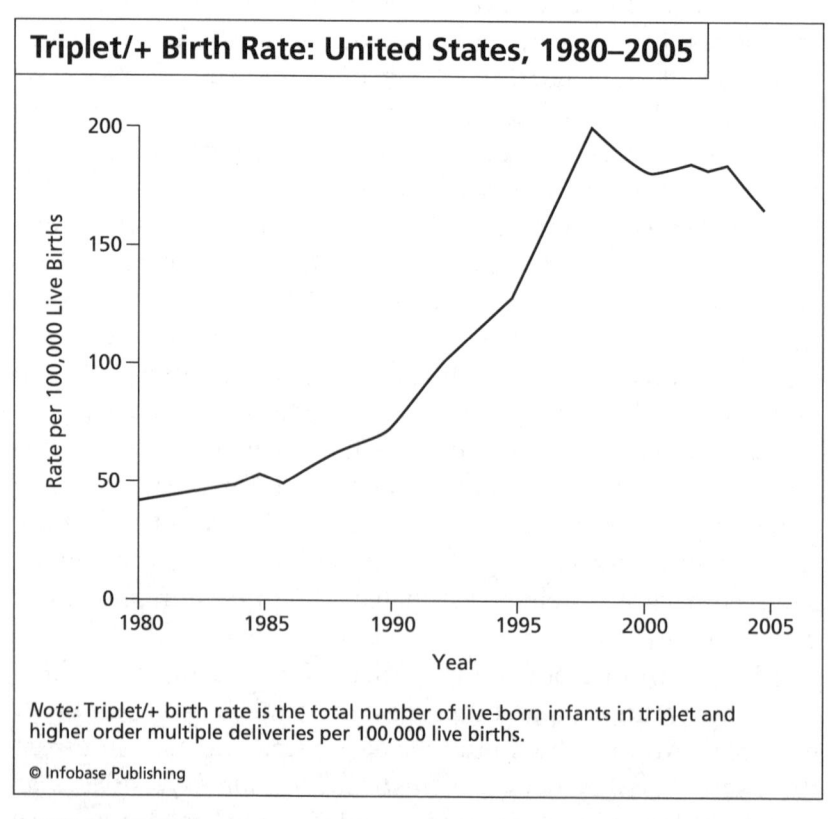

Triplet/+ Birth Rate: United States, 1980–2005

Note: Triplet/+ birth rate is the total number of live-born infants in triplet and higher order multiple deliveries per 100,000 live births.

© Infobase Publishing

Triplet/+ (triplet plus higher-order multiple) birth rate: United States, 1980–2005 *(Source: CDC National Vital Statistics Reports 56, no. 6, December 5, 2007)*

greater risk of miscarriage: An estimated 25 percent of women carrying quadruplets will miscarry in the first *trimester,* a risk that climbs to 50 percent for women carrying quintuplets.

There are also serious health risks for babies. Infants born in numbers greater than two often suffer debilitating health problems. "There's been a lot of concern about some of the serious consequences associated with the enormous rise in higher order multiple births over the past decade," CDC director Jeffrey Koplan said in 2001. "Most of these babies are born premature and of low birth weight, which puts them at risk for a variety of health threats, including infant death and severe life-long disabilities." Illnesses these children can suffer include heart and lung problems, *stroke,* blindness, developmental disabilities, and cerebral palsy.

The ethical principles of beneficence and respect for persons (see chapter 1) are central to the McCaugheys' story and others like it, where fertility drugs have been used in such a way as to threaten maternal and infant health. The principle of beneficence calls for health-care professionals to do their best to maximize benefit and minimize harm for their patients. Dr. Zev Rosenwaks, a fertility specialist with New York-Presbyterian Hospital, told the *New York Times* in June 1999, "A woman was not designed to have seven or eight children at one time. We are all very concerned that we practice medicine that does the least harm."

George Annas, chair of the Department of Health Law, Bioethics, and Human Rights at Boston University, concluded in no uncertain terms, "High-order multiples ought to be avoided. It's a preventable catastrophe."

The principle of respect for persons in this context requires that medical professionals do everything in their power to ensure that hopeful parents have been thoroughly informed of the real (and dire) risks associated with carrying multiple fetuses before they decide on a course of action. Medical ethicist Richard Zaner told the *New York Times* that obtaining true informed consent requires that "You do your level best to make sure that people understand—not that you've merely exposed the risk, but that they really understand what the risks are."

This is easier said than done says Sharon Covington, who works in psychological support services for people considering ART procedures. These patients, Covington notes, "are so overwhelmed with the whole process, they don't always hear."

Preventing Higher-Order Multiples

The McCaugheys rejected their physician's recommendation that they terminate some of the fetuses, since abortion would have gone against their religious beliefs. "God gave us those babies," Bobbi said. "He wants us to raise them." Kenny told reporters, "We were trusting in the Lord for the outcome."

Many medical ethicists focused their attention on the fact that a septuplet pregnancy was a preventable mistake—that had Bobbi's fertility treatments been managed responsibly, the parents never would have been faced with the difficult decision of whether to selectively reduce. The McCaugheys' fertility doctor stated at a press conference the day after the birth that she had administered the same dose of a fertility drug that had led to the birth of their daughter Mikayla. For reasons no one understood, "in this cycle we achieved more success than we could ever hope for."

Medical ethicists were unsatisfied with that explanation. "This multiple pregnancy simply did not have to happen," Arlene Judith Klotzko wrote in her 1998 article, "Medical Miracle or Medical Mischief? The Saga of the McCaughey Septuplets," published in the 1998 *Hastings Center Report*. "Good medical practice mandates ultrasound scans for women who have taken fertility drugs in order to monitor accurately the number of eggs they produce. If the number is too high and the risks of multiple pregnancy too great, the patient should be advised to refrain from sexual activity and try again later."

Critics also noted that in cases where a large number of eggs have been produced, ultrasound monitoring can determine when in vitro fertilization is a safer choice than sexual activity. Some of Bobbi's eggs could have been removed for fertilization and no more than three embryos implanted. Dr. Mark Sauer, then chief of reproductive endocrinology at Columbia Presbyterian

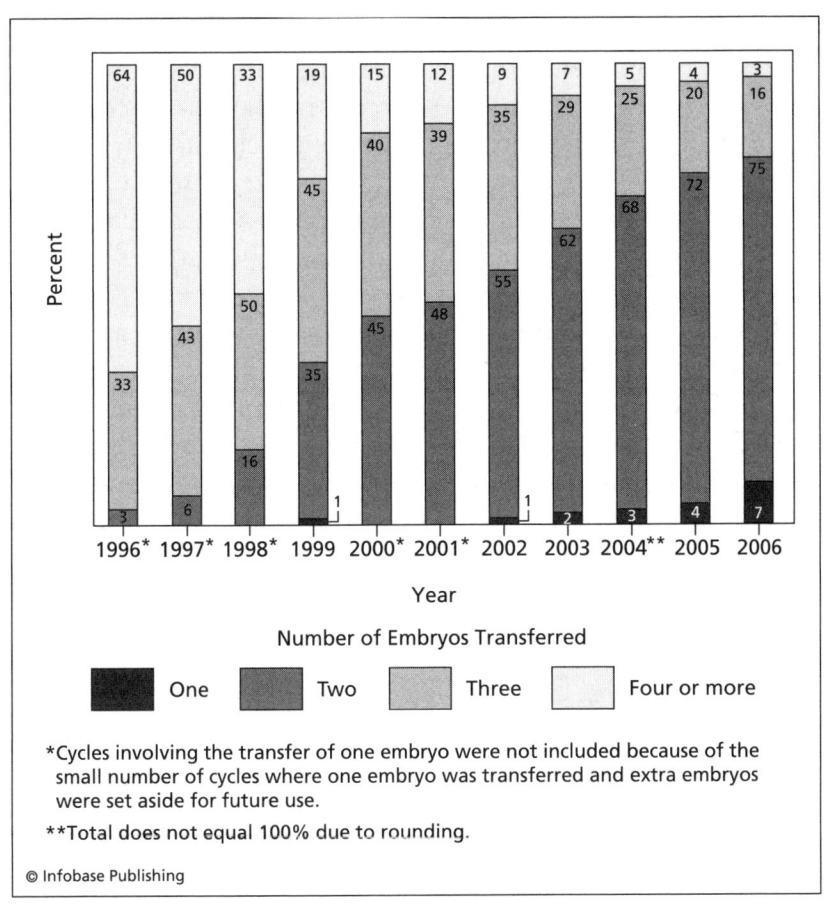

Percentages of fresh-nondonor cycles involving the transfer of one, two, three, or four or more embryos in women younger than 35 who set aside extra embryos for future use, 1996 to 2006 (*Source:* CDC, *2006 Assisted Reproductive Technology Report*)

Medical Center, told the *Washington Post* in November 1997, "it would have been obvious that her ovaries had overreacted to the drug. . . . She must have had a dozen or more eggs going, and if she was being monitored correctly they had to know she was grossly overstimulated before she got her HCG shot."

In cases where parents have already chosen IVF, the risk of a multiple pregnancy can be limited by implanting fewer embryos. The American Society for Reproductive Medicine currently recommends that two or more embryos be implanted in women over 35 (nearly 60 percent of all IVF procedures), but a growing number of reproductive specialists—in Europe in particular—have advocated *single embryo transfer* to protect maternal and infant health. Though pregnancy success rates can be lower with single embryo transfer, the dangers of multiple pregnancy are also greatly lessened.

In many European countries, fertility procedures are often subsidized, and therefore financial pressures to achieve success on the first try are greatly diminished. An American mother of IVF quadruplets, Marianne Jornlin, whose procedure cost $12,000, told the *New York Times* in June 1999, "Everything depended on that one cycle. The doctors gave me a 30 percent chance of a live birth. Where are you going to get the money for another cycle?" Mrs. Jornlin added, "I think now, going through all of this, if insurance had covered it, it would've been better to transfer two."

A 2007 study of single embryo transfer in women over 35 at the Stanford University School of Medicine found that success rates were surprisingly high; more than half of the women in the Stanford study (ages 35 to 43) became pregnant, as compared to a national success rate of about 25 percent in this age group. Stanford uses preimplantation genetic diagnosis (PGD) to determine which embryos are most likely to thrive before implanting them into a woman's uterus. (See chapter 3 for more on PGD.) "Although these results represent a selected group of patients," the Stanford investigators noted, "we believe that they should serve as encouragement to patients and providers who are considering single blastocyst transfer in the older IVF population."

SURROGATE MOTHERHOOD

Laws governing surrogate motherhood in the United States form a patchwork of conflicting rules and regulations, a legislative tangle that reflects the country's deep-seated ambiguity about the practice. On the one hand, there are strong emotional arguments in favor of surrogacy—a practice that is, in some cases, the only way for parents to have children who are their genetic offspring. These arguments often comingle with arguments in favor of freedom of choice for intended parents and surrogates, and they have prevailed in states where surrogacy is legal.

Ethical arguments against surrogacy can be grouped into three general categories, depending on the principles or beliefs underpinning them. First are arguments that surrogacy is another "unnatural" means of procreating (similar to arguments against IVF and other ART in general). These views are often grounded in strong religious beliefs. Then there are arguments based largely on the ethical principles of beneficence and respect for persons (see chapter 1). Critics who argue against surrogacy on these grounds assert that it is difficult or impossible to protect the interests of the surrogate or to fully inform her of the risks to her emotional health, since she cannot know in advance how difficult it might be to give up a child. Finally, there are arguments based on the principle of justice when a surrogate is paid for her services—especially if she is extremely underprivileged. Critics here argue that the practice of paying for surrogacy is exploitative—even coercive—of economically vulnerable women who, under less desperate circumstances, probably would not choose to take on the physical and emotional risks of surrogacy.

Twelve U.S. states refuse to recognize surrogacy contracts, but in the past five years, four new states—Texas, Illinois, Utah, and Florida—have joined the ranks of states explicitly legalizing and regulating the practice. Many hopeful parents from France, Germany, and other European countries where surrogacy is outlawed have formed agreements with surrogate mothers in U.S. states where the practice is legal.

Genetic Surrogacy: The Baby M Case
In March 1986, 29-year-old Mary Beth Whitehead gave birth to a baby girl who she had agreed to carry for William and Elizabeth Stern. Unlike most cases of surrogacy today, Mary Beth had donated the egg that resulted in the pregnancy (she

was artificially inseminated by William Stern's sperm), and so she was the baby's biological (as well as gestational) mother. After the birth, Mrs. Whitehead refused to surrender the baby—known to her as Sara Elizabeth and to the Sterns as Melissa Elizabeth ("Baby M" to the press)—and Mr. Stern sued to enforce the agreement and to strip Mrs. Whitehead of custody and visitation rights. The resulting conflict brought the new practice of surrogate motherhood to light and shaped the future of surrogate parenting in America.

A year after the baby's birth, a New Jersey judge upheld the surrogacy contract and denied Mrs. Whitehead parental rights. "The biological father pays the surrogate for her willingness to be impregnated and carry his child to term," Judge Sorkow said. "At birth, the father does not purchase the child. It is his own biological genetically related child. He cannot purchase what is already his." After handing down his opinion, Judge Sorkow summoned the Sterns to his chambers, where Mrs. Stern was immediately allowed to adopt the baby.

William Stern, biological father of Baby M, carries the infant from a visit with the biological and surrogate mother, Mary Beth Whitehead, during the first week of their custody trial. *(Bettmann/CORBIS)*

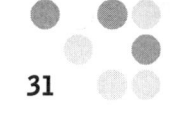

Feminist groups reacted angrily to the judge's opinion, accusing him of a clear bias in favor of the biological father over the biological mother. Noreen Connell, president of the New York State Organization of Women, told the *New York Times* after the ruling that the "mother was put on trial, and the father was not."

Surrogacy advocates, on the other hand, applauded the decision. "Surrogate parenting is here to stay," said the director of the Center for Surrogate Parenting in Los Angeles, William Handel. "It simply makes too much sense for too many infertile couples who have no other alternative." Though Handel and other surrogacy advocates were right that surrogacy was "here to stay," *genetic surrogacy* was not. The Baby M case changed the nature of surrogacy agreements in the United States, which now almost always involve the donation of an egg from a woman other than the surrogate.

Mrs. Whitehead appealed Judge Sorkow's decision, and nearly a year later, the New Jersey Supreme Court struck down the original surrogacy contract and restored Mrs. Whitehead's parental rights. Though the Sterns' residence would be the child's home, Mrs. Whitehead regained the right to visit the child.

When Melissa turned 18, she initiated proceedings to allow Mrs. Stern to adopt her legally. As a college student at George Washington University in Washington, D.C., she told *New Jersey Monthly* that it was strange to hear about the Baby M case in her medical ethics class. She was studying religion, and hoped to become a minister one day.

Gestational Surrogacy

Nearly all surrogacy agreements now require that the egg donor and the surrogate be different women, in order to avoid feelings of attachment between a surrogate and her biological offspring. Gernisha Myers, a gestational surrogate (GC) carrying twins for a couple in Germany, told *Newsweek* in March 2008 that when the agency asked her if she was afraid that she might get attached to the babies, she said, "'In a way, yes, even though I know they're not mine.' They said, 'Believe it or not, some GCs [gestational carriers] never feel any kind of bond.'

I found that hard to believe back then, but now I know what they're talking about. I don't feel that motherly bond. I feel more like a caring babysitter."

> ## SELECTIVE REDUCTION:
> ### When Surrogates and Intended Parents Disagree
>
> Surrogacy agreements often result in pregnancies with multiple fetuses, since the pregnancy is typically the result of an IVF procedure. In such cases, one medical option is to "selectively reduce" some of the fetuses in order to increase the chances that remaining fetuses will be born healthy. What if surrogates and intended parents disagree about the most ethical course of action?
>
> In 2001, Helen Beasley, a 26-year-old surrogate mother from England, sued a California couple for allegedly backing out of a surrogacy deal when they found out she was carrying twins. Ms. Beasley claimed that Charles Wheeler and Martha Berman, the intended parents, demanded that she abort one of the fetuses, while the lawyer for the couple said that it was more of a request—that the surrogacy contract had called for selective reduction if Ms. Beasley was found to be carrying more than one embryo.
>
> Ms. Beasley referred to a verbal agreement that selective reduction would happen before the 12-week mark and said that—although she notified them about the twin embryos at seven weeks—the couple made an appointment for her to undergo the procedure in week 13. She refused on health grounds, and, according to her lawyer, this was when Wheeler and Berman told her, "Well, we only wanted one. We don't want to separate them, so you figure out what you're going to do with the two babies."
>
> After the 14th week, according to Ms. Beasley, the couple stopped all contact with her. "As a surrogate," Ms. Beasley

Still, many GCs find it extremely difficult to give up a baby—sometimes much more difficult than they imagined. Another surrogate, Stephanie Scott, spoke of giving birth to a girl for a

> told BBC News, "I am supposed to get living expenses, lost wages, maternity clothes allowance—things like that. But I have not received a penny from this couple." In August 2001, Ms. Beasley sued the couple in San Diego Superior Court for breach of contract, fraud, and emotional distress. She also wanted Wheeler and Berman's parental rights revoked so that she could choose adoptive parents for the twins. Both she and the intended parents said that they had found other couples interested in raising them. Ms. Beasley already had a son and said that she could not afford to keep the babies.
>
> In Britain, the resolution would have been straightforward, since surrogate mothers retain legal rights to the babies for at least the first six weeks after birth whether or not they are genetically related to them. Under British law, Ms. Beasley would have had the right to place the babies with adoptive parents of her choosing. But the contract was signed in California—where Ms. Beasley was living before the birth—and in order to retain the right to choose the babies' fate, Beasley would need to prove that Wheeler and Berman were in breach of contract.
>
> The parties were unable to settle, and the case went to court. The resolution of the conflict is unknown, since the court's decision was sealed (kept private). How surrogacy disputes like this will be resolved in the future is anyone's guess. Ruth Claiborne, an adoption attorney, told CNN, "There is so little law on this. That's what makes the case so complicated and problematic.... I don't know of any cases that have addressed a situation like this one."
>
> Sanford Benardo, another adoption attorney, described the problem somewhat more bluntly: "We don't know who the parent is here."

couple on the opposite coast. "When she was born, they handed her to me for a second. I couldn't look, so I closed my eyes tight, counted 10 fingers and 10 toes, then gave her away. I cried for a month straight. I was devastated."

The bond that can develop, even with a genetically unrelated baby, was central to a famous 1990 court case in which Anna Johnson, a gestational surrogate, fought to retain custody of the baby she was carrying for Crispina and Mark Calvert. Mrs. Calvert had undergone a hysterectomy (an operation to remove her uterus) and was incapable of becoming pregnant, but her ovaries were unharmed. For a fee of $10,000, the Calverts entered into a contract with Ms. Johnson to carry an embryo, conceived by IVF, that was genetically Crispina's and Mark's. The embryo was implanted into Ms. Johnson's uterus and she became pregnant. In the seventh month of pregnancy, she changed her mind about giving up the baby and filed suit for custody, saying that she had bonded with the fetus. Her lawyer said, "Just because you donate a sperm and an egg doesn't make you a parent. Anna is not a machine, an incubator."

Ultimately a California Superior Court awarded the Calverts full custody and denied Johnson visitation rights. "I decline to split the child emotionally between two mothers," the judge ruled. He argued that while Johnson had fed and cared for the fetus as a foster parent would a child, she was a "genetic stranger" to the child and could not claim to be a parent based on her role as surrogate.

Motives for Surrogacy

A woman's motives for becoming a surrogate go directly to concerns about exploitation and coercion. Why would she choose such a physically and emotionally challenging path? Not surprisingly, surrogates' answers are as diverse as their personalities and life circumstances. Some women choose surrogate motherhood for economic reasons; others come to the decision largely because they want to give the gift of parenthood to people unable to carry a pregnancy. Many surrogates find that their motives are complex and hard to tease apart.

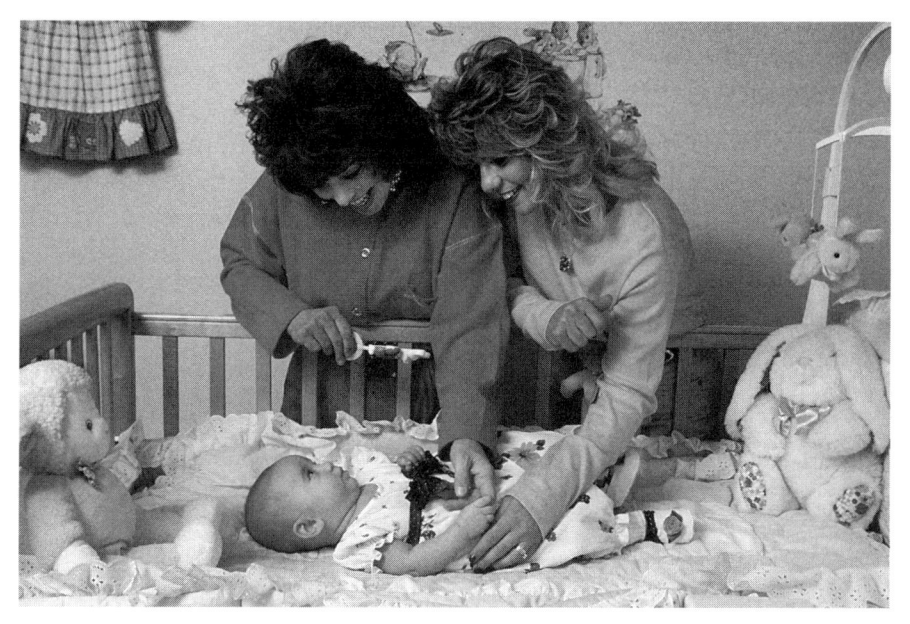

The woman on the right touches the four-month-old baby she bore as a surrogate mother on behalf of her sister, at left, who is the child's legal mother. *(Christopher Fitzgerald/The Image Works)*

Women who choose surrogacy partly for economic reasons are of special concern to medical ethicists, who highlight the potential to exploit low-income women. Some ethicists go so far as to argue that any commercial surrogacy agreement is exploitative and therefore should be outlawed—even when women do not make the decision based on economic need—because their altruistic, caring motives are being used to serve the ends of a commercial enterprise, and because the commodification of motherhood requires surrogates to repress or ignore potential parental feelings (to which a dollar value could never be assigned). Elizabeth S. Anderson argues in her paper, "Is Women's Labor a Commodity?" that "Commercial surrogacy constitutes a degrading and harmful traffic in children, violates the dignity of women, and subjects both children and women to a serious risk of exploitation. . . ."

Special concerns over the growing practice of hiring surrogates from low-income nations (particularly India) are considered later in this chapter, but the potential for exploitation is not limited to women from poorer countries. A surprising number of new American surrogates, for example, are military wives interested in supplementing their husbands' income. Women married to new enlistees can earn more from one surrogate pregnancy than their husbands earn in a year. Gernisha Myers told *Newsweek* in March 2008 that she chose surrogacy after the U.S. Navy transferred her husband to a different state and she was forced to leave her job as an X-ray technician. Once in their new home, she came across a local flyer with the ad, "Surrogate Mothers Wanted! Up to $20,000 Compensation!"

Most women who come to surrogacy in part—or even mainly—for monetary benefit also tend to rate the satisfaction from helping others as extremely rewarding. Surrogate Amber Boersma told *Newsweek*, "I felt like, 'What else am I going to do with my life that means so much?' . . . I thought I do not want to go through this life meaning nothing, and I want to do something substantial for someone else. I want to make a difference." Jennifer Cantor, another surrogate, described the experience this way: "Being a surrogate is like giving an organ transplant to someone, only before you die, and you actually get to see their joy."

Complex, interrelated motives can make it difficult to isolate potentially exploitative elements of surrogacy agreements, though in many cases, women would not choose surrogacy at all were it not for financial compensation. Some surrogacy agreements rely on economic inequalities more obviously than others; the growing business of surrogate motherhood in India is the subject of the next section.

SURROGACY ACROSS NATIONAL BORDERS

The practice of hiring surrogate mothers overseas—particularly low-income women from India—is growing steadily, and it is drawing fire from critics who say that it exploits an extremely vulnerable population of women who probably would not enter into a surrogacy agreement were they not living lives of extreme poverty. Hiring a surrogate in India costs on the order of $25,000

Controversies in Assisted Reproduction

(including medical bills and payments to the surrogate mother and clinic), whereas in the United States costs can range anywhere from $40,000 to $120,000.

Commercial surrogacy is illegal in twelve U.S. states and in several European countries; in France, for example, the highest court outlawed the practice in 1991, ruling that "The human body is not lent out, is not rented out, is not sold." Yet the market for surrogates has continued to grow, and India has stepped in to fill the void. Commercial surrogacy was legalized in that country in 2002, and now India does an estimated $445 million dollars a year of business in the "reproductive outsourcing" industry. There are no solid figures on how many babies are born each year to surrogate mothers in India, but some estimates place the number between 100 and 150, and growing.

Rudy Rupak, the head of a "medical tourism" agency based in California, told the *New York Times* in March 2008 that he expected to send at least 100 couples to India for surrogacy

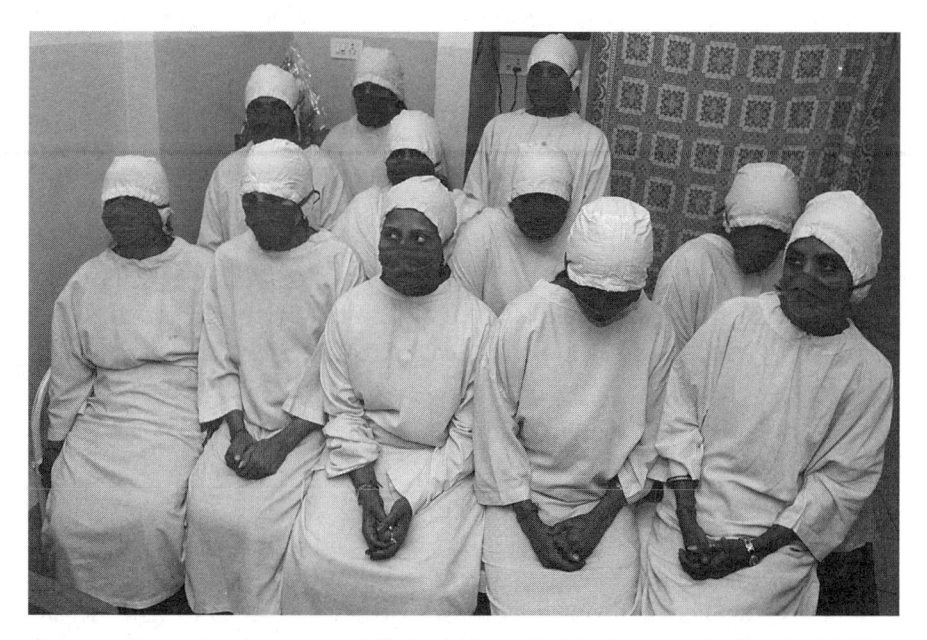

Surrogate mothers seen at Kaival Hospital in Anand, India, in 2006 *(AP Images/Ajit Solanki)*

by the end of the year. That number was up from 25 couples in 2007. "Every time there is a success story," Mr. Rupak said, "hundreds of inquiries follow."

Some supporters of international commercial surrogacy, including many Indian surrogate mothers, argue that the practice serves everyone's needs, providing babies for hopeful parents—most of them from the United States, Britain, and Taiwan—while also providing much needed money to low-income women. Surrogates in India are typically paid between $6,000 and $10,000, a fortune in a country where it might take 15 years to earn the same amount at a more traditional job. "From the money I earn as a surrogate mother I can buy a house," Nandani Patel told NPR's *Marketplace* via translator in December 2007. "It's not possible for my husband to earn more as he's not educated and only earns $50 a month, so nothing is saved." And Priyanka Sharma, a surrogate who is considering entering into a second contract, told NPR via translator, "Yes, I might do this again because after all there's nothing wrong in this. We give them a baby and they give us much-needed money. It's good for them and for us."

Then there are hopeful parents who point out that many people in more affluent countries would not be able to afford to hire a surrogate if it weren't for the less expensive overseas option. Lisa Switzer, a medical technician from San Antonio whose twins are being carried by a surrogate mother in India, told the *New York Times*, "Doctors, lawyers, accountants, they can afford it, but the rest of us—the teachers, the nurses, the secretaries—we can't, unless we go to India." Still, critics stress the extraordinary potential for exploitation of low-income women. Judith Warner, author of the book *Perfect Madness: Motherhood in the Age of Anxiety*, wrote in a January 2008 *New York Times* column that perhaps "when greater steps are taken toward improving international adoption procedures" and "when more substantive steps are taken to improve the health, status and education of women world-wide, it'll be easier to say with a clear conscience that what feels like callous exploitation really is just that."

SUMMARY

Surrogate motherhood and higher-order multiple births are two areas of special concern for maternal and infant health advocates, and cases like the McCaughey septuplets, the Beasley selective reduction lawsuit, and commercial surrogacy in India continue to put the principles of beneficence, autonomy, and justice to strenuous test in contexts where there may be several conflicting sets of "best interests" competing for priority.

On the subject of cross-cultural surrogacy contracts, Hilary Hanafin, chief psychologist at the Center for Surrogate Parenting, the oldest surrogacy agency in the country, told *Newsweek* in March 2008, "In what other world would you find a conservative [U.S.] military wife forming a close bond with a gay couple from Paris?" These new reproductive practices have brought about unprecedented social and cultural scenarios with ethical implications that are only beginning to unfold.

3

Eugenics, Genetic Testing, and Designer Babies

The past half-century has produced unprecedented expansion in scientific knowledge about the human *genome* and about the genomes (complete genetic information) of many animal, plant, and microbe species. The pace of discovery has been breathless: In 1953, James Watson and Francis Crick first described the double helix structure of *deoxyribonucleic acid (DNA),* the chemical substance that acts as a blueprint for building, running, and maintaining all living organisms. In April 2003—a mere 50 years later—sequencing of the human genome was complete.

This impressive surge in knowledge about our genes has been accompanied by intense hopes—and fears—about newfound technical powers to manipulate the production of life. This chapter will look at the current extent of our knowledge about the human genome; review the cruel history of *eugenics* (systematic genetic control) programs in the United States and other countries; and explore current implications of genetic discoveries for human health, privacy, and reproductive choice.

Codiscoverers of the structure of DNA, James Watson, at left, and Francis Crick, show their model of part of a DNA molecule in 1953. *(A. Barrington Brown/Photo Researchers, Inc.)*

GENETIC KNOWLEDGE, GENETIC CONTROL

Just a half-century ago, little was known about how inherited diseases can pass from one generation to the next. Some medical conditions have an obvious familial link, but the biological causes of inherited diseases were largely unknown until the 1970s, when researchers developed techniques to "read" the order of the chemical "letters" that make up the complex instructions contained in DNA. DNA is made up of two strands of nitrogen-rich molecules that are held together by weak hydrogen bonds. Each bonded pair (adenine together

Structure of DNA

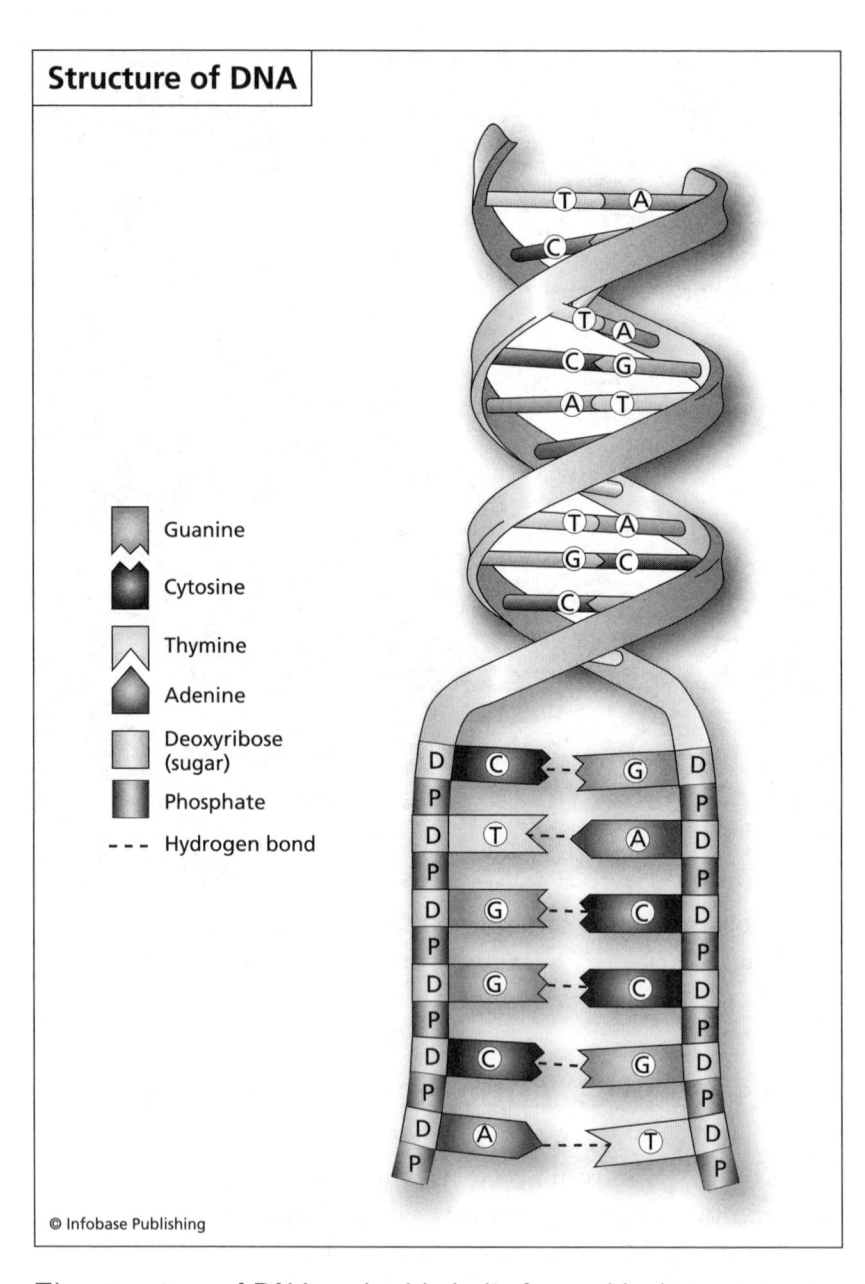

- Guanine
- Cytosine
- Thymine
- Adenine
- Deoxyribose (sugar)
- Phosphate
- --- Hydrogen bond

The structure of DNA, a double helix formed by base pairs attached to a sugar-phosphate backbone *(U.S. National Library of Medicine)*

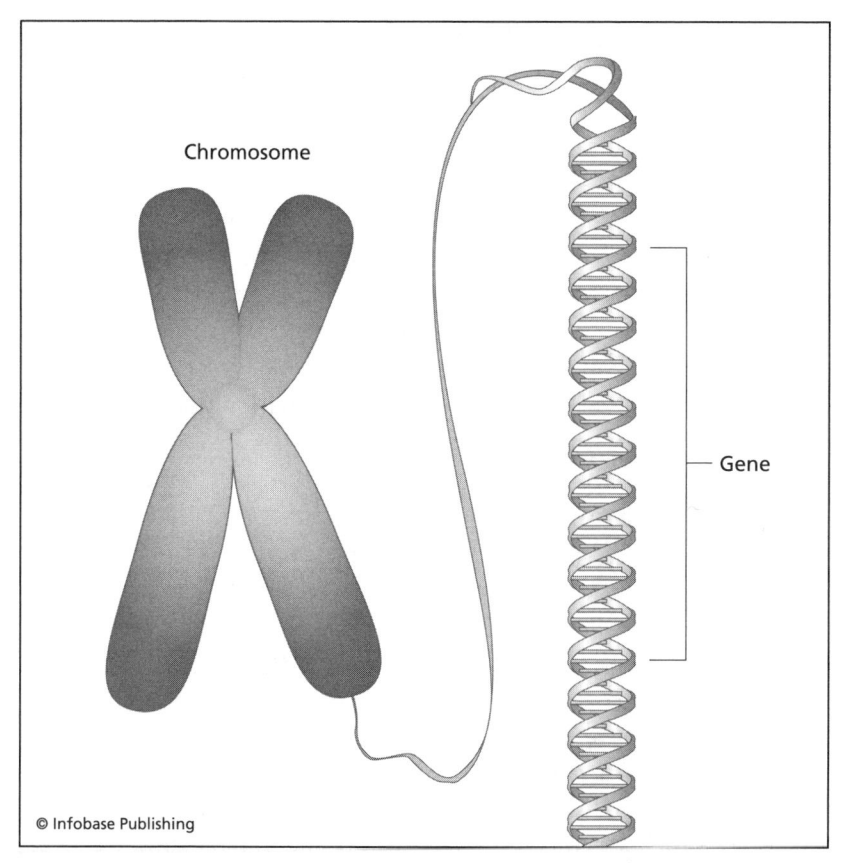

Eugenics, Genetic Testing, and Designer Babies 43

Genes and chromosomes. Genes are made up of DNA, and each chromosome contains many genes. *(U.S. National Library of Medicine)*

with thymine, AT, or guanine with cytosine, GC) is known as a *base pair*.

In 1990, the Human Genome Project—backed by the National Institutes of Health (NIH), the Department of Energy, and other government agencies around the world—undertook the sequencing of the approximately three billion base pairs in the human genome. The project identified more than 1,800 disease-causing genes and facilitated development of more than 1,000 tests for human conditions. According to NIH, there are more than 350 biotechnology-based drugs in clinical trials thanks to the project, and NIH hopes to cut the cost of sequencing an individual's genome to $1,000 or less.

More progress toward understanding the genetic basis of disease came in 2005 with the creation of HapMap, a resource that identifies and indexes genetic similarities and differences—variations known as *haplotypes*—in human beings, and then with the expansion and refinement of HapMap in 2007. This new data will make it easier to compare genetic differences between people with and without certain conditions so as to identify genetic factors involved in common human diseases like *heart disease, diabetes,* and *depressive disorders*. Unlike diseases with simple and direct genetic links (see the discussion of *Huntington's disease* at the end of this chapter), these conditions usually result from the combined effects of a number of genetic variations and environmental factors. HapMap has already been credited with finding genes involved in conditions like obesity and age-related blindness.

A new multinational initiative, The Cancer Genome Atlas (TCGA), was launched in the spring of 2007 and seeks to identify genetic variations seen in dozens of types of *cancer*. In their February 2007 *Scientific American* article, "Mapping the Cancer Genome," Frances Crick (codiscoverer of the double helix) and Anna D. Barker described hopes for the project, as well as myriad technical hurdles. "When applied to the 50 most common types of cancer, this effort could ultimately prove to be the equivalent of more than 10,000 Human Genome Projects in terms of the sheer volume of DNA to be sequenced."

A major challenge for investigators is distinguishing between rampant genetic "noise" in tumor samples and *mutations* that are cancer related. Researchers have found that there are sometimes large variations among tumor samples from patients diagnosed with the same type of cancer, supporting the idea that there are many genetic pathways from normal cells to cancerous ones. "As we survey the considerable empty spaces that exist in our current map of genomic knowledge about cancer," said Crick and Barker, "the prospect of filling those gaps is both exhilarating and daunting."

The human genome is not the only genome to be examined closely in recent years, and discoveries of disease-causing—

Eugenics, Genetic Testing, and Designer Babies   45

and disease-preventing—genes in many animal species have produced plenty of excitement. Tweaking a single gene in the roundworm *C. elegans,* for example, has been found to double

Tsila Levine is hugged by her biological mother, Margalit Omessi, near Tel Aviv, Israel, in August 1997 after genetic tests determined that they are mother and daughter, separated 49 years earlier when Tsila was allegedly snatched from her parents, who were Yemenite Jewish immigrants. Tsila later immigrated to the United States. *(AP Images/Barkai Wolfson)*

the worm's life span. Biogerontologists—researchers who study the biological mechanisms of aging—hope one day to apply that discovery to humans.

EUGENICS AS SOCIAL POLICY: THE CARRIE BUCK STORY

Enthusiasm about genetic science and its potential to improve human health, extend life spans, and alleviate suffering is prevalent in American culture, as is awareness of past attempts to "purify" the human species through genetic control—most notably, eugenics programs conducted throughout the first half of the 20th century in the United States and other countries. Eugenics, broadly speaking, is defined as any attempt to improve

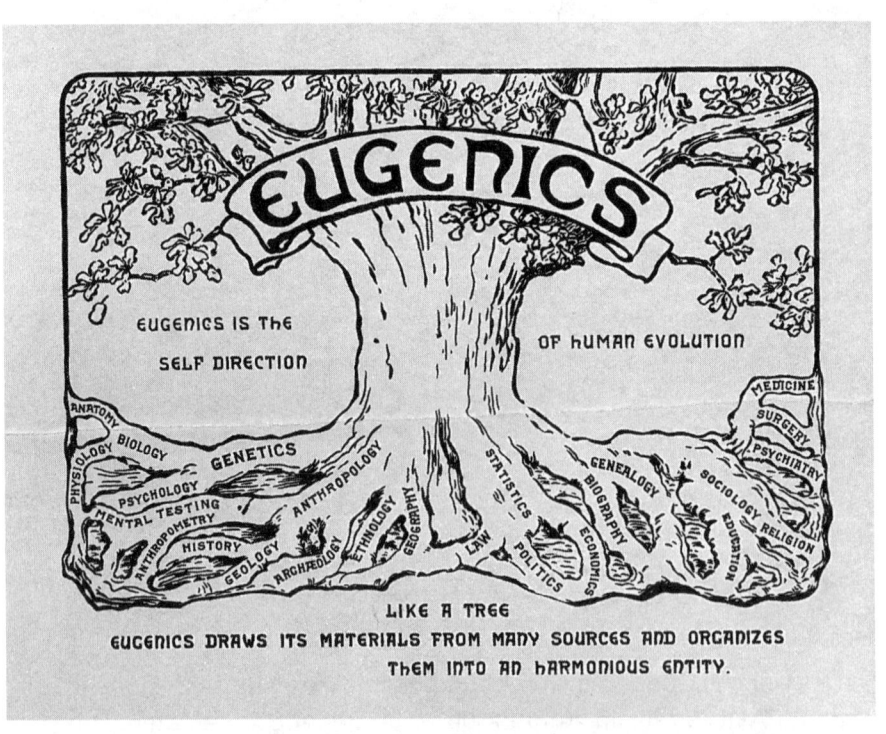

"Eugenics is the self direction of human evolution." Logo from the Second International Congress of Eugenics, 1921, depicting the new academic field as a tree fed by roots from a variety of disciplines
(American Philosophical Society)

the human species by genetic means, though historically the term has been applied to efforts to selectively breed people possessing "desirable" traits and restricting the reproductive rights of those possessing "undesirable" traits.

The word *eugenics* was coined in 1883 by the English scientist Francis Galton, cousin of Charles Darwin. Galton's first major work, *Hereditary Genius* (1869), advocated arranged marriages between men and women of distinction in the hopes of producing a generation of extraordinarily gifted British children. Galton wished to improve the human race by eliminating the "undesirables" and multiplying "desirables." The national eugenics programs of the early 20th century were closely aligned with this goal.

Nazi Germany's program of systematic sterilization is perhaps the most infamous example. Approximately 350,000 people were forcibly sterilized in that country in the 1930s, most of them for what was termed "congenital feeblemindedness," but some of them for blindness and deafness. Adolf Hitler declared in *Mein Kampf,* "The *völkisch* [populist] state must see to it that only the healthy beget children. . . . Here the state must act as the guardian of a millennial future. . . . It must put the most modern medical means in the service of this knowledge. It must declare unfit for propagation all who are in any way visibly sick or who have inherited a disease and can therefore pass it on."

Fascist Germany was hardly alone in its pursuit of genetic "perfection;" its sterilization program was modeled in part on the writings of American eugenicists. By January 1935, an estimated 20,000 people had been forcibly sterilized in the United States, approximately half of them in California. Virginia also pursued a zealous eugenics program against "mental defectives." That state passed its mandatory sterilization law in March 1924, when Carrie Buck was 18 years old, the mother of a young daughter, and an involuntary inmate of the Virginia State Colony for Epileptics and Feeble-Minded. She was the first person to be chosen for sterilization under the new law, and her story prompted a legal challenge against the constitutionality of forced sterilization—a case that went all the way to the U.S. Supreme Court.

Carrie and Emma Buck at the Virginia State Colony for Epileptics and Feeble-minded, Lynchburg, Va., ca. 1924. Carrie Buck was forcibly sterilized in 1927 under a state-run eugenics program. *(Arthur Estabrook Papers/University at Albany Libraries)*

Those in favor of sterilizing Ms. Buck argued that she was the second of three generations of women born "feeble minded"—that her mother and daughter both were mentally disabled, and therefore must be carriers of defective genes. Harry Laughlin, superintendent of the national Eugenics Record Office, said in a deposition for the case that "the evidence points strongly toward the feeble-mindedness and moral delinquency of Carrie Buck being due, primarily, to inheritance and not to environment."

The U.S. Supreme Court agreed, upholding in an 8-1 decision Virginia's sterilization bill and affirming the state's right to cut Carrie Buck's fallopian tubes. "It is better for all the world," wrote the famous jurist Oliver Wendell Holmes in the majority opinion on *Buck v. Bell*, "if instead of waiting to execute degenerate offspring for crime, or to let them starve for their imbecility, society can prevent those who are manifestly unfit from continuing their kind. The principle that sustains compulsory vaccination is broad enough to cover cutting the Fallopian tubes. Three generations of imbeciles are enough."

A 1985 review of the case by renowned Harvard professor of geology and zoology Stephen Jay Gould revealed that Carrie Buck was not, in fact, institutionalized for a lack of intelligence, but instead for an illegitimate pregnancy resulting from rape by a family acquaintance—what Laughlin had characterized as Ms. Buck's "moral delinquency." It was common practice at the time to send women away to institutions to hide embarrassing pregnancies.

Carrie Buck, in fact, was of apparently normal intelligence, as was her daughter. A leading scholar of the court case, Paul A. Lombardo of the University of Virginia School of Law, saw her some years after her forced sterilization. "As for Carrie," he wrote in a letter to Gould, "when I met her she was reading newspapers daily and joining a more literate friend to assist at regular bouts with the crossword puzzles. She was not a sophisticated woman, and lacked social graces, but mental health professionals who examined her later in life confirmed my impressions that she was neither mentally ill nor retarded."

In 1980, Dr. K. Ray Nelson, then director of the hospital where Ms. Buck had been sterilized, discovered records of more than 4,000 sterilizations, the last as late as 1972. Sifting through those files, he uncovered the terrible fact that Carrie Buck's sister Doris, who had always wanted a child, was sterilized without her knowledge. She had been told the operation was for appendicitis.

CONTEMPORARY LIBERAL EUGENICS

Holmes's rhetoric about Carrie Buck—and others he deemed "unfit" to procreate—seems shockingly insensitive less than a century later. Some advocates for minority and disability rights, though, are concerned that even today, newfound genetic knowledge—much like the surgical knowledge employed in the Carrie Buck case—might be used to "fix" genetic traits in people who do not themselves see these traits as defective or undesirable.

National eugenics programs are hardly historical artifacts. In the People's Republic of China, for example, the Maternal and Infant Health Care Law (in effect since 1995) was designed, in the words of the Minister of Public Health, to actively "prevent new births of inferior quality." The law discourages marriage and pregnancy in cases where a potential parent has a hereditary condition, or a condition that may or may not be due to genetic causes, such as mental illness, retardation, or learning disabilities.

Policies restricting people's reproductive rights strike most Americans as distasteful, but a new, more liberal form of eugenics has emerged in wealthy countries like the United States, where many people have access to IVF and preimplantation genetic diagnosis (PGD). This liberal, free-market form of eugenics is framed in terms of providing choices and protecting reproductive freedoms—rather than placing restrictions on them. It has engendered much less controversy than state-imposed eugenics programs, though many religious leaders believe that PGD—even for serious diseases—is morally wrong for the same reasons that they believe IVF is morally wrong.

Screening Embryos during IVF

For more than a decade, prospective parents undergoing IVF have used preimplantation genetic diagnosis to screen embryos for fatal childhood diseases like *Tay-Sachs,* a heartbreaking genetic disorder that causes accumulation of a fatty substance in the nerve cells of the brain. PGD is now routinely used to screen for other serious diseases like spina bifida, cystic fibrosis (CF), *muscular dystrophy, hemophilia,* and *sickle-cell disease.*

A growing number of parents are choosing to screen embryos for milder diseases or for diseases that may never develop. PGD is now available for many conditions that strike later in life or have high cure rates, like colon cancer; for more manageable diseases, like *arthritis*; and for conditions that have less than a 50 percent chance of developing, like many adult-onset cancers.

A 2006 survey of fertility clinics conducted by the Genetics and Public Policy Center indicated that among clinics that offer PGD, 28 percent have used it to detect genes for diseases that do not strike until adulthood. In a controversial new practice, some

Eugenics, Genetic Testing, and Designer Babies

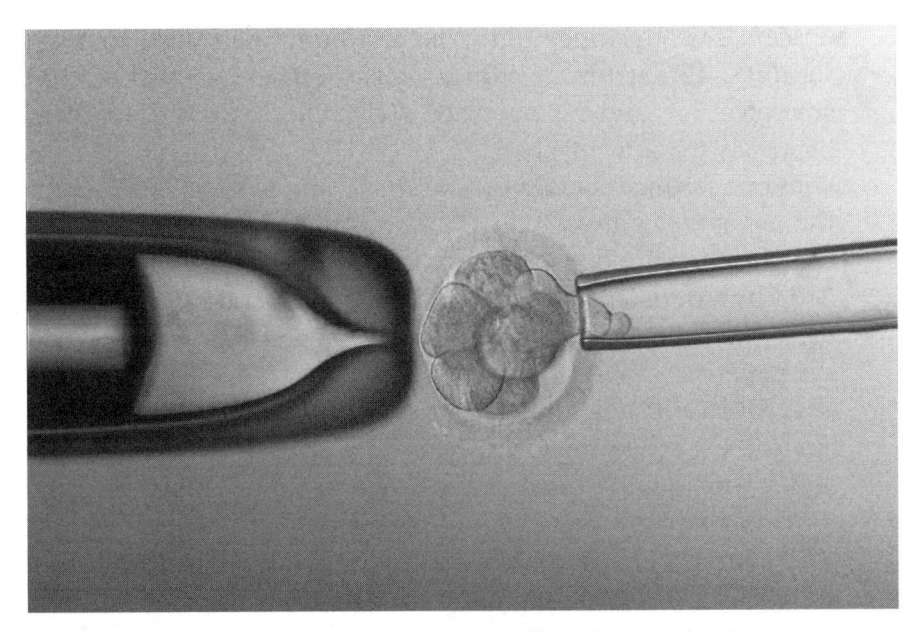

Preimplantation genetic diagnosis (PGD). A pipette *(at left)* holds an eight-celled human embryo *(near center)* produced by IVF. A smaller pipette *(at right)* draws off one cell from the embryo after its membrane has been punctured with acid. The cell is then genetically screened to check for disorders like Down syndrome. If found to be normal, this embryo will be implanted in the womb. *(Pascal Goetgheluck/ Photo Researchers, Inc.)*

parents with conditions like deafness and dwarfism have used PGD to increase their chances of having children who share their condition. According to the Genetics and Public Policy Center, 3 percent of American clinics offering PGD have used it "to select an embryo for the presence of a disability." Pediatric cardiologist Darshak M. Sanghavi commented in a December 2006 essay in the *New York Times,* "It turns out that some mothers and fathers don't view certain genetic conditions as disabilities but as a way to enter into a rich, shared culture."

The use of PGD to select a male or female embryo raises eyebrows, but it is becoming more common in the United States. In 2004, when the Genetics and Public Policy Center asked Americans if they approved of using PGD for sex selection, 40 percent said that they did. And fertility clinics are listening: In 2006, 42 percent of clinics that offered PGD also offered sex selection, and 9 percent of PGD reportedly is performed expressly for this purpose.

Some fertility doctors believe that using PGD to test just for the sex of the embryo is unethical, but that if embryos are being tested for medical reasons, and the parents already have a child and want to "balance" the family with a child of the opposite sex, there is no good reason to withhold the genetic information. "It's the patient's information, their desire," Dr. Jamie Grifo of New York University's Fertility Center told the *New York Times* in February 2007. "Who are we to decide, to play God? I've got news for you, it's not going to change the gender balance in the world."

Other doctors say they will perform PGD for sex selection with a first child. "We prefer to do it for family balancing," clinic owner Dr. Jeffrey M. Steinberg told the *Times,* "but we've never turned away someone who came in and said, 'I want my first to be a boy or a girl.' If they all said a boy first, we'd probably shy away, but it's 50-50."

Much of the concern over the use of PGD for sex selection comes from the Chinese and Indian experiences, where it has become more common to abort female fetuses now that ultrasound and other tests can identify sex in utero. Chinese officials noted that in 2005, 118 boys were born for every 100 girls. In 2001, the Indian Supreme Court ordered strict adherence to a ban on prenatal gender screening (in place since 1994), but critics argue that the government has been lax in enforcing the ban, and that no physician has ever been convicted of *sex-selective abortion.*

Many ethicists hold that PGD for certain very serious diseases—like Tay-Sachs, spina bifida, or muscular dystrophy—is justifiable, since in these cases it will prevent needless and serious suffering. Another argument for PGD is that it is preferable to *selective abortion* of a fetus after prenatal testing detects the potential for disease or deformity. Says pediatrician and ethicist Jeffrey R. Botkin, "PGD is ethically permissible for its primary purpose, that is, to offer couples at high risk of bearing a child with a significant genetic condition the opportunity to have a healthy child without resorting to selective abortion."

But what counts as significant? As PGD becomes more commonplace, that line has begun to shift. "Reproductive choice, as far as I'm concerned, is a very personal issue," says Dr. Steinberg of his patients' wishes for PGD. "If it's not going to hurt anyone, we go ahead and give them what they want."

Screening Donors for IVF

A growing number of prospective parents handpick eggs or sperm from donors who have been screened to match certain desired traits, such as race, family health history, even eye color. Hopeful parents can pay a fee that allows them to search donor pools for a wide range of characteristics. Says commentator David Brooks of this relatively new practice, "At this very moment thousands of people are surfing the Web looking for genetic material so their children will look nothing like me . . . These sites take sex and turn it into shopping. They allow you to browse through page after page of donor profiles, comparing weight, noses, personality and what one site calls 'tannability.'"

A woman in San Antonio, Texas, made headlines in January 2007 for taking the commodification of assisted reproduction one step further. Jennalee Ryan offers a service selling embryos created from eggs and sperm from donors who have been pre-screened for good physical and mental health, education level, and physical appearance. Besides medical and psychological tests and genetic screenings, Ms. Ryan requires at least five color photos of the donor and of the donor's children and siblings (if any) before she will approve them. In this way, Ms. Ryan says that prospective parents can "choose a donor with similar characteristics" to their own.

"It's tempting to label Ryan a madwoman, as many critics have," wrote columnist William Saletan in January 2007. "But that's exactly wrong. Ryan represents the next wave of industrial rationality."

Prenatal Screening

Long before PGD was an option, prenatal screening presented many parents with the choice to abort fetuses with serious

(continues on page 56)

INFANT SCREENING FOR TREATABLE CONDITIONS

The March of Dimes recommends that all newborns receive screening tests for 29 disorders for which effective treatment is available. Most of these conditions are inherited, and they can be grouped into the following five broad categories:

Amino acid metabolism disorders. These inherited conditions are caused by deficiencies in particular enzymes critical to healthy metabolic functioning. A lack of these enzymes can lead either to toxic levels of *amino acids* (the building blocks of proteins) or of ammonia (a by-product of protein breakdown) in the body. *Phenylketonuria (PKU),* for example, is a rare but serious condition that affects more than one in 25,000 newborns and can result in toxic levels of the essential amino acid phenylalanine in the bloodstream.

With PKU—as with many of the disorders in this category—newborns can suffer severe mental retardation unless the condition is detected and treated early. If newborns with PKU are kept on a diet low in the amino acid until the age of six, the severity of retardation can be greatly reduced.

PKU screening was the first large-scale infant screening program mandated by state laws, and it is generally considered the most successful. In 1962, Dr. Robert Guthrie developed an easy, inexpensive procedure for testing infants' blood for the disease, and Massachusetts quickly passed the first mandatory testing law. Within four years, PKU testing was required in 41 states.

Organic acid metabolism disorders. These inherited conditions result from inactivity of an enzyme involved in the breakdown of amino acids and other important organic substances in the body (lipids, sugars, and steroids), which leads to toxic levels of these chemicals in body tissues. Treatment for most of these conditions includes a low-protein diet and nutritional supplements.

Fatty acid oxidation disorders. In this category of inherited disorders, enzymes required to break fat down into energy in the form of glucose (sugar) do not work properly. When cells run out of energy—especially if a person skips meals—coma or even death can result. Treatment usually includes avoidance of fasting and nutritional supplements.

Hemoglobinopathies. These are inherited disorders of the red (oxygen-carrying) blood cells due to abnormal types, or amounts, of the protein *hemoglobin.* This group of conditions includes sickle-cell disease, in which abnormal hemoglobin causes red blood cells to be stiff and abnormally shaped. The effects of this disorder can vary greatly from person to person, ranging from almost no ill effects at all to pain and organ damage, even stroke and death. Young children with the disorder are especially vulnerable to dangerous infections, and should receive all standard childhood vaccinations.

Screening for sickle-cell disease has a controversial history in the United States. The disorder is most prevalent among African Americans, though it can also affect people of Mediterranean, Caribbean, and Central and South American ancestry. Early efforts to screen for the disease resulted in misguided state laws that required African Americans applying for marriage licenses to undergo testing (since the disease only occurs when a child inherits a gene from both parents). Some insurance companies began to require that African-American employees be screened, and sometimes people who carried the gene were denied jobs. The U.S. Air Force dismissed 143 African Americans simply because they carried the trait, even though none of them actually had the condition. The Air Force eventually withdrew its testing requirement when a trainee filed a lawsuit.

In April 1993, an expert panel administered by the Public Health Service recommended screening all newborns—regardless of race—for sickle-cell anemia—the most common form of sickle-cell disease—and currently more than 40 states screen for the disorder.

(continues)

(continued)

Others. This catchall category includes all remaining conditions (some of which are not necessarily heritable) for which early diagnosis can alleviate the effects of the disorder. Congenital *hypothyroidism,* hearing loss, and cystic fibrosis (CF) are by far the most common conditions in this group, each with an incidence greater than one in 5,000.

Treating hypothyroidism with oral doses of thyroid hormone can prevent brain damage in infants and allow for normal growth throughout childhood, while early detection of hearing problems can allow for the use of hearing aids and may prevent serious speech and language deficits. Some studies show that early diagnosis and treatment of CF can improve the growth of babies and children, though CF remains a very serious condition with no cure.

Newborn screening requirements by state can be viewed at the March of Dimes Web site (www.marchofdimes.com/peristats).

(continued from page 53)
conditions. Ultrasound, maternal blood screening, *chorionic villus sampling (CVS),* and *amniocentesis* can be used to detect—with varying degrees of accuracy—the presence of a chromosomal abnormality (such as *Down syndrome* or *trisomy 18*), or genetic or other birth defects.

Selective abortion of fetuses on the basis of diagnoses like Down syndrome has provoked much more controversy than abortion of fetuses with fatal conditions like Tay-Sachs or *anencephaly* (severely underdeveloped brain and skull). Down syndrome can be accompanied by physical problems like heart defects, but many children with the syndrome lead apparently happy, meaningful lives. Nevertheless, raising a severely impaired child can bring hardship and social isolation on parents and family members; for this reason, many people believe that aborting fetuses diagnosed with Down syndrome and other similar impairments is justified. (See chapter 8 for the ongoing debate over appropriate treatment of impaired infants.)

GENETIC DISCRIMINATION AND PRIVACY

In April 1999, Terri Seargent began to have trouble breathing. She made an appointment to see her doctor, and the results of a simple genetic test came back positive for alpha-1 antitrypsin (AAT) deficiency, the same respiratory disease that killed her brother. AAT is a protein produced in the liver that protects against enzymes capable of destroying lung tissue. If detected early, AAT deficiency is treatable with replacement therapy. The test saved Ms. Seargent's life, but when her employer learned of her potentially expensive medical condition, she lost her job and her health insurance.

Ms. Seargent was not alone in her plight. A survey of more than 1,500 genetic counselors and doctors conducted by University of Massachusetts Medical Center social scientist Dorothy C. Wertz found that 785 patients reported losing jobs or insurance due to results of genetic tests. An earlier study conducted by Georgetown University found that 13 percent of patients said that they had been refused or fired from a job because of a genetic condition, and a 1999 survey by the American Management Association found that 30 percent of large and midsize employers were seeking genetic information about their employees. Seven percent of those companies said that they used that information in hiring and promotion decisions.

According to the National Human Genome Research Institute's (NHGRI's) fact sheet on genetic discrimination, "Public fears about genetic discrimination mean that many individuals do not participate in important biomedical research at the NIH. Many patients also refuse genetic diagnostic tests that help doctors identify and treat diseases: they worry that they will lose their health insurance if it is proven that they are genetically predisposed to a disease." Barbara Fuller, a senior policy advisor at NHGRI, told *Scientific American* in January 2001 that one-third of women contacted for possible inclusion in a breast cancer study refused to participate out of fear of genetic discrimination.

NHGRI, along with many lawmakers and patients' advocates, has fought for legislation to protect patients from genetic discrimination in the workplace. Since the late 1990s, most states have passed laws protecting the genetic privacy of workers (see the Web site for the National Conference of State Legislatures for a detailed summary), and in May 2008—13 years after genetic privacy legislation was first introduced in the U.S. House of Representatives—Congress passed the federal Genetic Information Nondiscrimination Act (GINA). Under GINA, employers can be fined as much as $300,000 for using genetic information in hiring, firing, or salary decisions, and insurance companies are no longer permitted to use genetic information to deny individuals benefits or raise their insurance premiums.

"This clears away what in many people's mind had been a real cloud on the horizon," said Dr. Francis Collins, director of NHGRI. "Families with a strong history of genetic disease will have one less worry about the circumstances they find themselves in, and hooray for that."

SCREENING CHILDREN AND ADULTS FOR INCURABLE DISEASES

Katharine Moser was just 23 when she learned that she carried the gene for Huntington's disease, an incurable condition that begins in middle age and slowly destroys the cells of the brain. Ms. Moser's grandfather had suffered the ravages of the disease for three decades, and she was one of a small—but growing—minority of young people who choose to be tested.

When the genetic counselor told her that the test had come back positive, Ms. Moser asked, "What do I do now?"

"What do you want to do?"

"Cry."

Ms. Moser did not regret being tested—she had wanted to know—but she had believed deep down that the test would come back negative. Now she had to find a way to live with the fact that it had not.

"I'm going to become super-strong and super-balanced," she told her best friend. "So when I start to lose it I'll be a little closer to normal." By "lose it," Ms. Moser was referring

Eugenics, Genetic Testing, and Designer Babies   59

Autosomal Dominant Genetic Traits

Affected father

Unaffected mother

Affected son

Unaffected daughter

Unaffected son

Affected daughter

■ Unaffected   □ Affected

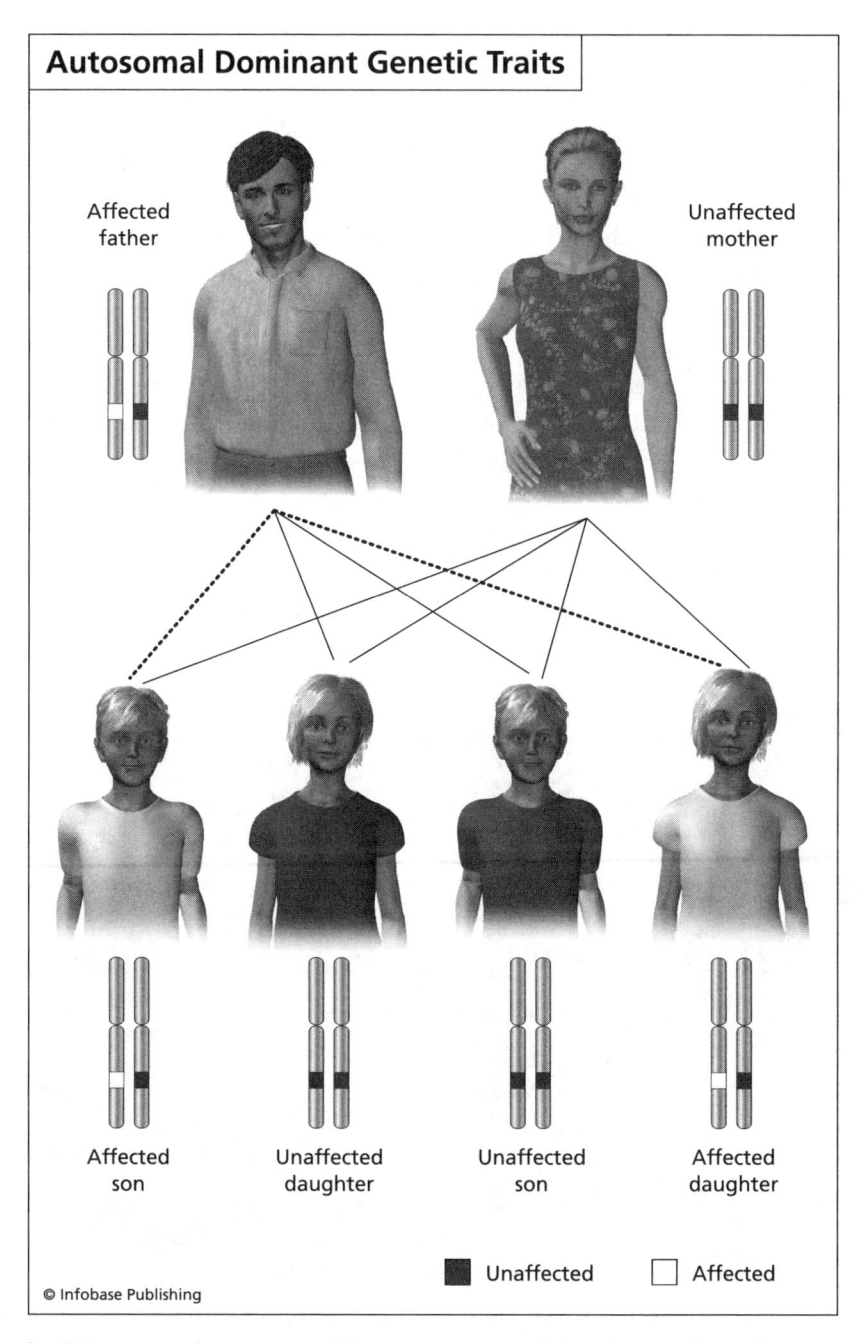

© Infobase Publishing

In this example, a man with an autosomal dominant disorder has two affected children and two unaffected children. *(U.S. National Library of Medicine)*

to the debilitating progression of the disease, which typically sets in between the ages of 30 and 50 and can begin with mild personality and mood changes or uncontrollable twitching and jerking. Eventually, Huntington's robs people of the ability to walk, talk, swallow, and think. Memory and judgment become severely impaired, and some patients become paranoid, manic, or violent. Death is the eventual result, usually through damage done to the brain, though suicide is not uncommon.

A vast spectrum of genetic disorders exists, some painful and lethal, others less serious or curable. Some genes are statistically associated with a disease, but do not automatically lead to its occurrence. The genes BRCA1 and BRCA2 are examples; they are associated with increased risks for breast and ovarian cancers, but they do not always result in illness.

Having the gene for Huntington's, by contrast, means that a person will one day develop the disease, and as of the early

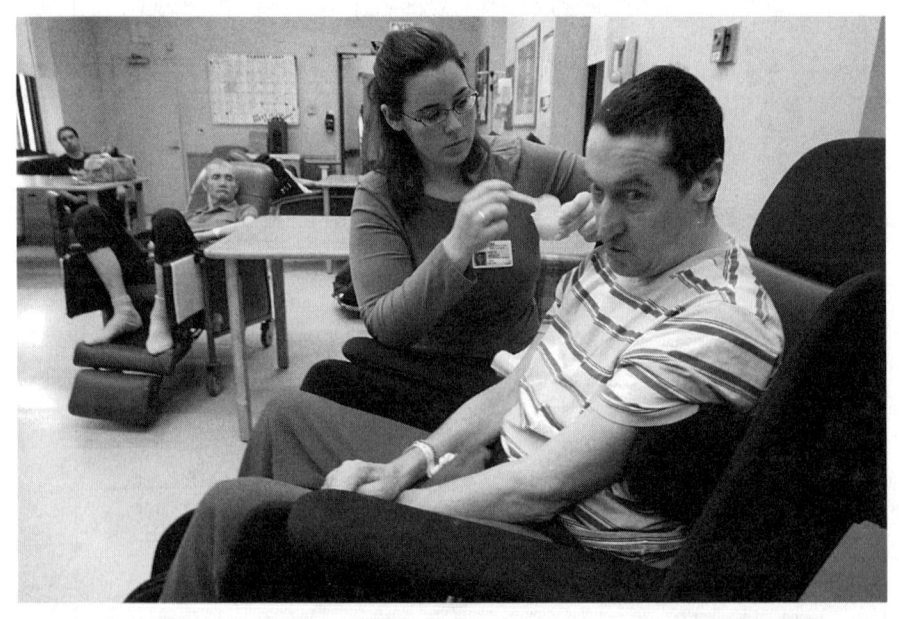

Katharine Moser, an occupational therapist, feeds a patient with Huntington's disease. Moser was diagnosed with Huntington's at the age of 23 after choosing to have a genetic test for the condition. *(Suzanne DeChillo/The New York Times/Redux)*

21st century, this disease is incurable. Huntington's is what is known as an *autosomal dominant disease,* meaning that it is caused by a single defective gene on a nonsex chromosome. It does not require combination with another defective gene in order to be expressed (as do genes for recessive diseases, such as hemophilia), and it cannot be prevented by social, biological, psychological, or other environmental factors.

Until recently, most young people with Huntington's in their family history have chosen not to find out whether they carry the gene. Even if a parent has already succumbed to the disease, there is a 50 percent chance that the child is not a carrier. When people do find out that they have the gene, they sometimes battle depression, and genetic counselors are trained to caution young people about the potential psychological impact of a positive result.

"We're seeing a shift," Dr. Michael Hayden, a professor of human genetics at the University of British Columbia, told the *New York Times* when it reported Ms. Moser's story in March 2007. "Younger people are coming for testing now, people in their 20s and early 30s; before, that was very rare. I've counseled some of them. They feel it is part of their heritage and that it is possible to lead a life that's not defined by this gene."

The genetic test is no crystal ball. There are still significant unknowns, including how particular individuals will react to the painful news. "What runs in your own family, and would you want to know?" Nancy Wexler, neuropsychologist and president of the Hereditary Disease Foundation, told the *Times*. "Soon everyone is going to have an option like this. You make the decision to test, you have to live with the consequences."

SUMMARY

In August 1978, *U.S. News & World Report* ran a quote by British Labor MP Leo Abse warning that the birth of Louise Brown foreshadowed "a time when an embryo could be sold, guaranteed free of genetic defect and in which the color of eyes, sex, and probably size on maturity could be specified."

Three decades later, his words seem prescient. Though preimplantation genetic diagnosis and prenatal screening are no guarantee against disease, prospective parents can choose these tests to screen for genetic predispositions for particular defects or conditions—even, in the case of PGD, to select embryos for implantation on the basis of traits like sex or the presence of a disability.

Once a baby is born, screening is recommended for a range of genetic conditions with effective treatments. An increasing number of older children and adults are also choosing to be tested for genes that predispose them to serious diseases—some of which, like Huntington's and *Alzheimer's disease,* have no known cure.

Genetic researchers are investigating a host of biochemical techniques that show promise for correcting disease-related genes. This research is the subject of the next chapter.

Gene Therapy and Enhancement

This chapter describes the array of biochemical techniques known as gene therapy and tells stories of past failures and recent successes in human trials, drawing on a range of perspectives—from those with high hopes that the research will save millions of lives, to those primarily concerned with its potential dangers.

The last section of the chapter looks at the current debate in academic and policy circles over what counts as therapy and what counts as enhancement, and over who—as *genetic enhancements* become available—should have the power to alter their genes or the genes of their offspring. But first, a brief account of how the techniques work.

WHAT IS GENE THERAPY?

Genes are composed of specific sequences of base pairs that "spell out" instructions for the production of proteins essential to life. Mutated (altered) genes often receive attention for causing genetic disease, but it is the absence or faulty operation of basic proteins that generate symptoms associated with particular genetic conditions. When genes are mutated in such a way as to leave important proteins absent, in short supply, or unable to function properly, normal biochemical operations that build and maintain living organisms can be severely disrupted.

Gene therapy refers to a collection of techniques designed to correct or replace a gene that is not operating normally. Until 2006, no gene therapy trials in humans had met with unequivocal success. Some had failed quite tragically, resulting in disease and death for some participants. No human gene therapy is currently available outside of clinical trials, but recent limited successes in three human experiments—one to treat *myeloid blood diseases,* one to treat advanced skin cancer, and one to treat inherited childhood blindness—have given researchers new hope that gene therapy may hold profound potential to save many lives, and soon.

Techniques to Correct Faulty Genes

Researchers may follow one of several lines of attack to correct a mutated gene:

1. *Insertion.* The most common approach is to insert a normal gene somewhere along the genome. The normal gene begins to produce whatever protein has been lacking from the nonfunctional gene.
2. *Replacement.* In some cases, an abnormal gene may be "swapped" for a normal gene through an exchange of material between strands of DNA (a process called *homologous recombination*).
3. *Repair.* In other cases, a nonfunctional gene may be repaired by correcting the specific point where the harmful mutation occurs. This technique is known as *reverse mutation.*
4. *Regulation.* A gene may be turned "on" or "off" depending on whether it is helpful in fighting disease or whether it is one of its causes.

Gene Therapy Delivery Systems

How is new genetic information introduced into the cells of animals or humans in a gene therapy experiment? Researchers are currently investigating four major delivery systems:

Gene Therapy and Enhancement 65

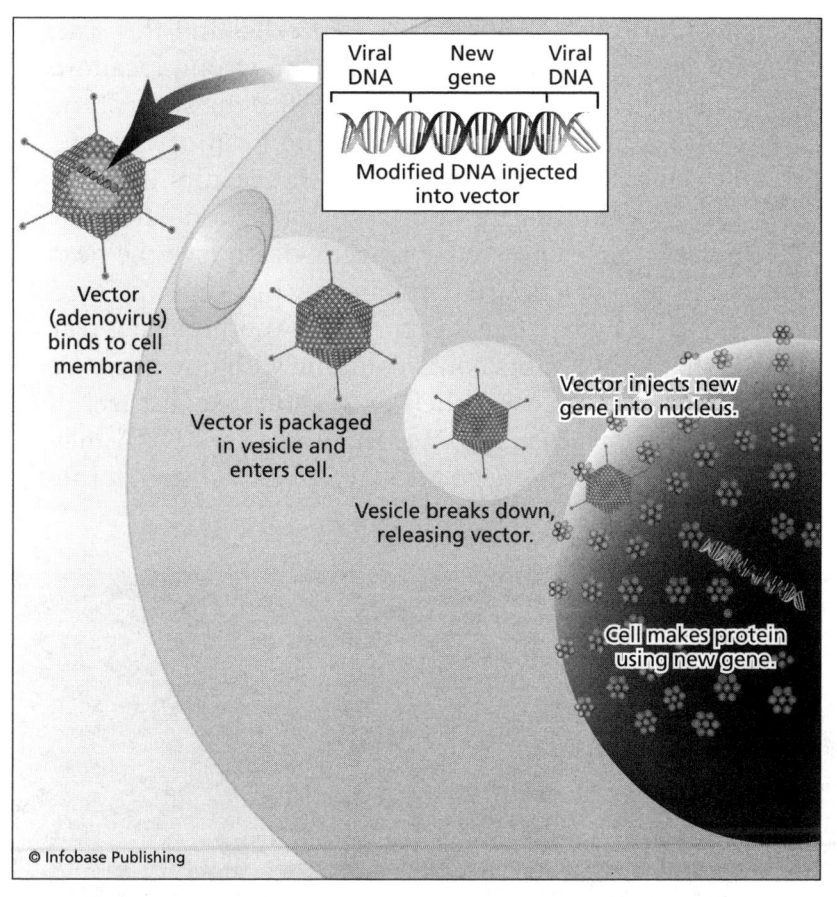

Viral DNA | New gene | Viral DNA
Modified DNA injected into vector

Vector (adenovirus) binds to cell membrane.

Vector is packaged in vesicle and enters cell.

Vesicle breaks down, releasing vector.

Vector injects new gene into nucleus.

Cell makes protein using new gene.

© Infobase Publishing

Gene therapy using an adenovirus vector. A new gene is injected into an adenovirus vector, which is then used to introduce the new DNA into a cell. If the treatment is successful, the new gene will start producing a functional protein.

1. *Viral vectors.* Different types of viruses (including *retroviruses, adenoviruses* like the common cold virus, and herpes viruses) can be engineered in such a way as to inactivate their disease-causing genes while allowing them to carry therapeutic genes into cells of the body. Although viruses are the most common delivery method in gene therapy trials to date, there are serious risks associated with viral vectors. Viruses can infect

healthy cells as well as diseased cells, and they can insert themselves into DNA in the wrong location, causing cancer or other diseases. (See the story of the *SCID* gene therapy trial under the heading, "Status of Gene Therapy Research.") Viral vectors can also cause an extreme immune system reaction—the suspected cause of Jesse Gelsinger's death (see the next section)—or a desired protein may be over-expressed to such a degree that it is harmful. Another major fear is that viral vectors could combine with other genetic material and recover their power to cause disease.

2. *Direct introduction of DNA.* In some cases DNA may be directly introduced into target cells, although this

Parkinson's sufferer Nathan Klein, standing with his family, addresses a press conference announcing the first gene therapy trial for Parkinson's disease in August 2003. *(AP Images/Stuart Ramson)*

technique is uncommon, since it only works with certain tissues.
3. *Artificial lipid spheres.* These fatty particles can pass therapeutic DNA through the membranes of target cells, though they tend to be a less efficient delivery system than viruses.
4. *Molecular vectors.* Recent laboratory experiments have transferred DNA to cells via artificial, biodegradable polymers (large organic molecules), stoking hopes that some molecular vectors may be as efficient as viral delivery systems, while presenting fewer risks to patients.

THE DEATH OF JESSE GELSINGER

On September 9, 1999, 18-year-old Jesse Gelsinger caught a plane from Tucson to Philadelphia and admitted himself to the hospital at the University of Pennsylvania. He had consented to participate in a gene-therapy trial with the Institute for Human Gene Therapy to test a technique designed to supply the gene for the enzyme *ornithine transcarbamylase (OTC)*. A lack of OTC causes a severe metabolic disorder in which ammonia can accumulate to dangerously high levels in the bloodstream. Jesse had lived with the disease since at least the age of two, when he fell into a level-one coma and was diagnosed. A combination of a low-protein diet and medications kept his condition stable for the majority of his young life.

OTC deficiency is a rare disorder occurring in one out of 40,000 newborns. Most infants born with the genetic mutation usually become comatose and die within 72 hours of birth. Jesse's case was relatively mild, since he was a genetic hybrid—also known as a chimera—with a mixture of healthy and abnormal cells in his body. Since there was no family history of the disease, researchers speculated that his condition was probably caused by a spontaneous mutation in the gene that produces OTC.

When Jesse consented to take part in the study, he understood that he could not hope to gain therapeutic benefits from his participation. The technique was not expected to cure the disease, but instead to alleviate it temporarily so as to protect infants' brains from toxic damage. (Ethical concerns with testing the procedure on sick infants led Arthur Caplan, medical ethicist at University of Pennsylvania, to conclude that the only acceptable test subjects would be carriers of the disease or people like Jesse, who live with a very mild form of the deficiency.)

Jesse knew that even if the genes were taken up by his cells and began producing OTC, any benefits would be short lived. The technique used a weakened strain of adenovirus (the virus that causes colds) to deliver the normal gene into Jesse's liver cells. His immune system was expected to make quick work of the virus.

It was explained to Jesse that *hepatitis* (liver inflammation) might result from the procedure, but it is unclear whether he was told that three monkeys had died of liver inflammation and a blood-clotting problem when they were given a stronger strain and higher dose of the adenovirus.

On the morning of September 13, Jesse received an injection of the genetically altered adenovirus, making him the 18th participant in the study. He received the highest dose administered in the experiment, though the previous participant had received the same dose and had tolerated it well.

Jesse, tragically, did not. By that night, he had a dangerously high fever and was showing signs of jaundice. He developed a blood-clotting problem similar to the one that proved fatal to monkeys after they received the stronger strain of the virus, and his blood ammonia spiked. He slipped into a coma. His father, Paul, arrived from Tucson on September 15, and Jesse was placed on a ventilator for breathing problems. By that night, it was clear that his lungs were failing. Jesse's doctors put him on a machine that removed carbon dioxide from his blood.

"If we could just buy his lungs a day or two," Dr. Steven Raper later described the physicians' thinking to the *New York Times Magazine*, "maybe he would go ahead and heal up." But Jesse did not heal. He suffered multiple organ failure and irreversible brain damage. On the morning of September 17, doctors asked for Paul's permission to turn off the ventilator.

After a brief service, Paul addressed a crowd of family and staff in the room before giving the signal to stop the machine. "Jesse was a hero," he told everyone.

At 2:30 P.M. on September 17, Jesse Gelsinger was pronounced dead.

The Aftermath

A high-profile investigation followed, and the FDA found that the Institute for Human Gene Therapy had behaved improperly by not reporting liver toxicity in four patients prior to Jesse, or the deaths of monkeys injected with a similar vector. Dr. James Wilson, head of the institute, protested that he had reported the liver toxicity information to the FDA and that the monkeys had died from injection with a stronger virus.

Paul Gelsinger felt that he and Jesse had been misled by researchers' emphasis on success in animal trials, and he stated before a meeting of the federal Recombinant-DNA Advisory Committee (RAC) that he had been made to believe that gene therapy had been successful in humans. He learned at that meeting that Dr. Wilson and the University of Pennsylvania were major stockholders in Genovo, a pharmaceutical company holding rights to develop any of Wilson's clinical products for commercial sale, and that Wilson had sold his 30 percent share for $13.5 million.

Following Jesse's death, the FDA put a temporary halt to two similar gene-therapy trials in humans. The exact biological causes of Jesse's death may never be determined with certainty, though the presence of abnormal cells in his bone marrow led investigators to suspect that he suffered a severe immunological attack on the experimental viral vector.

STATUS OF GENE THERAPY RESEARCH

Jesse's death was a major blow to those with high hopes for gene therapy. The Institute for Human Gene Therapy was restricted by the FDA to basic research, and it ceased to exist the following year. Other human research proceeded cautiously until 2002, when the field suffered another major setback: Two children in Paris developed a leukemia-like condition after receiving

Toddler Wilco Conradi, nearly three years old in this photo, visits the aquarium at the Artis Royal Zoo in Amsterdam in August 2002. The family outing would have once been impossible, as Wilco was born with severe combined immunodeficiency (SCID), or "bubble boy" disease, and had to be isolated in a plastic enclosure to protect him from infections. An experimental gene therapy restored his immune system and gave him a normal life. *(AP Images/Peter Dejong)*

experimental gene therapy for SCID (*severe combined immune deficiency,* or "bubble boy" disease).

Before this disheartening result, the SCID study had seemed an unqualified success. Researchers had used a genetically engineered retrovirus to insert a new, healthy gene to correct a defect on the *X chromosome*. Most children born with the faulty gene die within their first year, but eight of the 11 patients in the trial were considered cured after administration of the healthy gene.

Sadly, something went wrong for two patients. Later studies revealed that the healthy gene, which was carried by the retroviral vector, had inserted itself too close to a cancer-related gene and activated it, resulting in the children's blood disease.

After learning of the problem, the FDA temporarily halted all gene-therapy trials using retroviral vectors in blood stem cells. In April 2003 it eased the ban, allowing similar gene-therapy trials to proceed for the treatment of life-threatening diseases.

In April 2008, researchers in the United Kingdom announced a successful gene therapy trial in a patient with a type of inherited blindness called *Leber's congenital amaurosis*. (The new technique was used in a total of three patients, one of whom showed significant improvement in his night vision.) The condition, which is caused by a defect in a single gene, appears at birth or in the first few months of life and prevents the retina from detecting light correctly. It causes progressive deterioration in eyesight and has no known treatment. The trial, conducted by Moorfields Eye Hospital and University College London, inserted healthy copies of the gene into patients' retinal cells using a viral vector. The three participants showed no apparent side effects, and researchers hope to achieve better results in younger patients.

Retroviral vectors have been used with limited success in two other recent human trials, one to treat myeloid blood diseases and one to treat advanced *metastatic melanoma* (a deadly form of skin cancer). In the melanoma study, patients' own white blood cells were engineered to become cancer-fighting cells. "These very exciting successes in treating advanced melanoma bring hope that this type of gene therapy, altering lymphocytes, could be used in many types of common cancers and could be achievable in the near future," said acting director of the National Cancer Institute, John E. Niederhuber.

TECHNICAL LIMITATIONS AND ETHICAL CONCERNS

Despite recent achievements in the field, serious technical challenges and ethical dilemmas have made successful human trials the exception, not the rule. Gene therapy, says Dr. David Williams, director of experimental hematology at Cincinnati Children's Hospital (one of the institutions involved in analyzing the myeloid disease study), "will succeed in treating some devastating genetic illnesses in children" and "is just beginning to fulfill some of the predictions made in the 1980s and early 1990s," but Williams and others caution that there are many unknowns and that the technology will need ongoing improvement.

Some of the technical hurdles researchers hope to overcome include the following:

1. *The need for multiple rounds of therapy.* The challenges of introducing therapeutic DNA into the genome, along with the rapid division rate of many cells, often make multiple rounds of gene therapy necessary to achieve long-term benefits.
2. *The body's immune response.* The immune system often learns to attack the gene therapy as it would any foreign substance, rendering it less effective.
3. *Dangers associated with viral vectors.* See a full discussion of the risks of viral vectors in the section, "What Is Gene Therapy?"
4. *Disorders involving multiple genetic and environmental factors.* Many common diseases like diabetes and heart disease do not result from a single genetic factor but from multiple genetic and environmental causes. Such diseases will prove more difficult to treat with gene therapy than single-factor genetic diseases.

These technical challenges are inseparable from ethical concerns, since weighing harms and benefits requires accounting for unknown risks—both to patients consenting to participate in the research, and in some cases, to future generations whose genetic makeup could be altered in deleterious ways.

Problems with Informed Consent— The Case of Jolee Mohr

When researchers are unsure of the effects that newly introduced genes—or the vectors used to deliver them—will have on the human body, how can a patient make a truly informed decision? The patient needs to understand that disease and death are real possibilities. The patient also must understand that

Gene Therapy and Enhancement

most preliminary gene-therapy trials are not expected to result in therapeutic benefit for research subjects themselves.

In the face of unknown risks, ethical concerns about human trials can be grouped under two ethical principles—respect for persons and beneficence. Respect for persons means that an individual's autonomy—or ability to make independent choices—will be preserved through the process of informed consent, which is taken to include three basic elements: information, comprehension, and voluntariness. The principle of beneficence requires that researchers work hard to minimize possible risks and maximize possible benefits to research subjects—in other words, that investigators will do their absolute best to ensure that subjects' interests are served. (See chapter 1 for a more detailed description of these ethical principles.)

Robb Mohr cries during an interview at home in August 2007. Robb's wife, Jolee, died of a severe fungal infection three weeks after receiving experimental gene therapy for arthritis. The cause of the deadly infection is still unknown. Photos of Jolee and their young daughter, Toree, hang on the wall. *(AP Images/Seth Perlman)*

A 2007 tragedy—the death of a young woman possibly linked to gene therapy—illustrates the extreme challenges of obtaining valid informed consent in gene therapy trials. Though the death may never be blamed definitively on the trial, a close look at the story reveals missteps that the *Washington Post* called "failures in the safety net that is supposed to protect people" from the risks of gene therapy.

Jolee Mohr, 36, was a healthy woman who experienced occasional stiffness from arthritis. Her husband reported that she had never missed work due to her condition, but she was recruited by her personal rheumatologist for a study to test the safety of a new gene therapy for arthritis. Dozens of other patients had already received injections of the gene without major side effects, but three weeks after her second injection of the experimental gene therapy with a viral vector, Ms. Mohr died of multiple organ failure and internal bleeding.

An autopsy found that a fungal infection—one that normally would cause only mild illness—had gone out of control. The question remains whether the injections suppressed her immune system to the point where the fungus could take over. The most suggestive finding was that the viral vector and its active gene, which was supposed to produce an anti-inflammatory protein in Ms. Mohr's knee, had spread to other organs.

"The biggest question I have is would my wife still be alive today if she hadn't participated in this study?" her husband said in a statement before an NIH review committee. "I have it in my heart that she'd still be here." Ms. Mohr also left behind a five-year-old daughter.

The situation is complicated by the fact that Ms. Mohr, in addition to receiving the experimental treatment, was taking three other immune-suppressing drugs for her arthritis. The drugs had not caused problems prior to the gene therapy injections. When NIH initially approved human trials to begin testing the therapy, the experimental protocol involved only a single dose to patients who were not receiving other drugs. Some members of the NIH review committee expressed serious reservations about administering the experimental therapy to patients on immune-suppressing drugs, and some wondered whether the risk of even one shot was worth it for non-life-threatening diseases like arthritis.

Since that initial trial, the approval process for so-called follow-on studies shifted to the Food and Drug Administration (FDA) and has been held behind closed doors ever since, making it impossible for the public to scrutinize the decision-making process that led to approval of this, and many other, human gene-therapy trials built on earlier studies.

Medical ethicists point to two guidelines of clinical research that were broken when Ms. Mohr signed up for the study. First, she should have been required to take consent forms home for review rather than signing them on the spot. Second, if a patient's personal doctor is heading a study, someone else should be in charge of describing the study to the patient.

"Because of the relationship . . . you have to worry that they won't listen carefully enough to the risks." Hank Greely, director of the Center for Law and the Biosciences at Stanford, told the *Washington Post* that patients might be tempted to think, "'After all, if my doctor is doing this, it must be good for me.' That can be difficult to overcome with words in a consent form."

Jonathan Moreno, medical ethicist at the University of Pennsylvania, said that real clinical possibilities were too buried in the thick technical language of the consent form. Patients could not be expected to understand the risks as they were written. "Even a smart person would have a very hard time figuring out what they're talking about," said Moreno.

A longtime criticism of the FDA review process for gene therapy is that the agency is prohibited by law from publicly reporting so-called adverse events in clinical trials if they involve what a pharmaceutical company regards as proprietary, or trade secret, information. In this way, investigators conducting similar trials may have no knowledge of adverse patient reactions and may not be able to fully inform their own subjects.

Critics see this restriction on FDA authority as wholly incompatible with the principle behind informed consent. "The law," says medical ethicist Ronald Munson, "seems to favor protecting the investments of the pharmaceutical industry more than the protection of human subjects."

Moreover, since 2000, companies have been granted up to a year to report adverse events to the FDA, provided that the lead investigator does not think that the test product is responsible for the problem. Serious events linked to experimental treatment, on the other hand, are to be reported to the FDA "as soon as possible" and no later than seven days following investigators' knowledge of the complications.

Targeted Genetics, the biotech company sponsoring the arthritis trial, first classified Jolee Mohr's problems as not serious, and then as serious, but not related to gene therapy. No report was made to the FDA and the trial continued, until Ms. Mohr needed emergency transport to the hospital at the University of Chicago, two-and-a-half weeks after she began having problems. At this point Targeted Genetics submitted a "serious adverse event" report to the FDA.

Following Ms. Mohr's death, Adil Shamoo, molecular biologist at the University of Maryland and editor of the journal *Accountability in Research,* told the *Washington Post,* "There are no uniform standards for 'adverse events' reporting. And there is no motivation to report them. . . . No one wants to show their dirty linen."

Germ-Line Therapy

Human gene therapy research has, so far, been limited to changes in an individual's genome. The potential exists, however, to alter genetic material carried by egg and sperm and thus to change the genetic makeup of future generations. Experiments affecting offspring have already been carried out in animal species; pigs, for example, have been genetically engineered to produce offspring with organs that might be better tolerated by human beings.

Since eggs and sperm are the seeds—or so-called germ cells—of future generations, this type of genetic change is known as *germ-line therapy*. It would, according to the National Cancer Institute (NCI) fact sheet on gene therapy, "forever change the genetic makeup of an individual's descendents. Thus, the human gene pool would be permanently affected. Although these changes would presumably be for the better, an error in

Named ANDi (backwards for "inserted DNA"), this rhesus monkey received an extra bit of genetic material to become the world's first genetically modified primate. Because the DNA was slipped into his mother's egg prior to fertilization and could be passed onto future generations, the procedure falls into the category of germ-line engineering. *(Oregon Regional Primate Research Center, the Oregon Health Sciences University, Portland, Oregon)*

technology or judgment could have far-reaching consequences. The NIH does not approve germ-line therapy in humans."

Though no germ-line gene-therapy trials have been approved, NCI acknowledges there is a risk that gene therapy trials could affect an individual's germ line by mistake. When viruses or liposomes are used to deliver DNA, "there is a slight chance that this DNA could unintentionally be introduced into the patient's reproductive cells. If this happens, it could produce changes that may be passed on if a patient has children after treatment."

Some ethicists argue that the risks are not large enough to outweigh the potential benefits of germ-line research. "The diseases are so serious," say medical ethicists Ronald Munson and Lawrence Davis, "and the promise of the therapy so great, that it would be wrong to give into the objections that have been raised to gene therapy. If they are allowed to prevail, then the social and scientific support needed to realize the therapeutic possibilities of gene therapy may never materialize. This outcome would be as wrong and almost as serious as if we had failed to develop and use antibiotics or vaccines."

Therapy or Enhancement?

No present-day debate challenges mainstream ideas about health and disease more provocatively than the debate over what counts as therapy and what counts as enhancement, and who should have access to the latter. A genetic change blocking a cancer-causing gene once a tumor has started to grow would count as therapy, of course, but what about a genetic intervention designed to enhance the immune system of a healthy person, or even to improve the mathematical or verbal aptitude of a person of average intelligence?

This section looks at the debate between mainstream physicians and academics on one side, and transhumanist thinkers on the other, about where legitimate health-care needs stop and human "enhancement" begins. The medical possibilities can sound like the stuff of science fiction, but many interventions with the potential to drastically alter human life spans and abilities are in the works. The ethical debate hinges on when, if ever, a society should deny its citizens medical enhancements they feel will improve their lives; and whether, if some people have access to expensive genetic interventions, those interventions should be guaranteed to everyone who wants them.

Medical ethicist Eric Parens observes in his 2005 *Hastings Center Report* essay, "Authenticity and Ambivalence: Toward Understanding the Enhancement Debate," that many people will feel morally ambivalent when they reflect on some of the technological choices available to them. "If understanding is what we are after," he says, "we should embrace rather than

THE TRANSHUMANIST MOVEMENT

A growing number of people consider themselves part of a new movement called *transhumanism,* which seeks, in part, to challenge mainstream cultural assumptions about what should count as normal. Subscribers to transhumanism want to know why steroids for athletes are banned while muscle-enhancing surgery is permitted, and why children are allowed to drink caffeine but not take other performance-enhancing stimulants.

The World Transhumanist Association describes itself as a nonprofit group that "advocates the ethical use of technology to expand human capacities." The organization supports "the development of and access to new technologies that enable everyone to enjoy *better minds, better bodies* and *better lives.* In other words, we want people to be *better than well.*" Specifically, the organization supports the improvement of the human species (and in some cases, other species) through the use of "present technologies, such as genetic engineering, information technology, and pharmaceuticals, as well as anticipated future capabilities, such as nanotechnology, machine intelligence, uploading, and space colonization."

"Transhumanists," according to their association, "advocate the moral right for those who so wish to use technology to extend their mental and physical (including reproductive) capacities and to improve their control over their own lives." Gene therapy should not be limited to people with disabilities or diseases, transhumanists believe, but to anyone wishing to improve their physical or mental functioning.

In its 2003 report, "Beyond Therapy: Biotechnology and the Pursuit of Happiness," President George W. Bush's bioethics council recommended against using technology to give people powers they do not have naturally. But human enhancement is happening already—some military pilots use the stimulant drug Modafinil, for example—and an increasing number of medical ethicists want to sit down with transhumanists and talk about it.

(continues)

> *(continued)*
>
> In May 2006—the same weekend that the blockbuster super-mutant movie *X-Men: The Last Stand* hit movie screens—Stanford University's Center for Law and the Biosciences hosted a conference on human enhancement and whether all humans should have a right to it. Walter Truett Anderson, then president of the World Academy of Art and Science, delivered the keynote address, asking participants to look at the implications of human enhancement in a global context. Anderson and many others believe that if human life spans are extended the way transhumanists wish them to be, the natural environment might be taxed beyond repair.
>
> The time has arrived, says Anderson, for serious talk about the risks and benefits of technologically enhancing humans. "There are a lot of issues that are going to begin to surface," he told MSNBC. "People will have to confront them."

suppress the ambivalence we often experience when we think about specific interventions."

Gene therapy is an intervention many people feel morally ambivalent about, but a growing number of Americans find the idea of genetically altering themselves or their children to be an acceptable—even praiseworthy—moral choice. A 2003 Virginia Commonwealth University survey found that 41 percent of respondents thought that changing a baby's genetic characteristics to reduce the risk of serious disease was an appropriate use of medical technology, and 37 percent of respondents said they would be likely to alter their own genes if it meant slowing the aging process.

In light of the astonishing pace at which genetic knowledge is being acquired, the question is not if, but when, genetic enhancements to our mental and physical abilities will become available. When they do, who should have access to them? Should they be considered a luxury or a basic human right? If considered a luxury, inequalities in society will be exacerbated, and new ones might emerge. If considered a basic human right, who should pay for them?

The National Cancer Institute frames potential problems with genetic enhancement this way: "[T]here is concern that such manipulation could become a luxury available only to the rich and powerful. Some also fear that widespread use of this technology could lead to new definitions of 'normal' that would exclude individuals who are, for example, of merely average intelligence."

SUMMARY

When Steptoe and Edwards conceived a child under laboratory conditions, onlookers expressed wonder and fear at the doors their work might open to genetic control of babies at the embryonic stage. In August 1978, *U.S. News & World Report* noted that IVF's success set the stage for "the screening of eggs, sperms and embryos for defects," while *Newsweek* emphasized how new genetic techniques might one day "be used to alter the genes of human fetuses just fertilized in the test tube." Three decades later, gene therapy in children and adults foreshadows the serious potential for genetically engineered fetuses and germ-line therapy.

In its 1994 report, "Ethical Issues Related to Prenatal Genetic Testing," the Council on Ethical and Judicial Affairs of the American Medical Association concluded that manipulation of genetic material "to alter benign characteristics or traits should be approached with extreme reservation," and that in general "such manipulation is inappropriate, and its use should be strongly discouraged." In such "exceptional" cases in which altering non-disease traits or characteristics might be acceptable, the AMA laid down three minimal ethical criteria: "there would have to be a clear and meaningful benefit to the child; there could be no trade-off with other characteristics or traits; and all citizens would have to have equal access to the genetic technology, irrespective of income or other socioeconomic characteristics."

Whether these ethical lines will hold if gene therapy saves more lives than it risks—and if cultural attitudes toward it continue to soften—is anyone's guess.

5

Stem Cells and Therapeutic Cloning

Stem cell research has provoked widespread public debate since 1998, when scientists first isolated stem cells from human embryos and grew them in the laboratory. Under the right conditions, embryonic stem cells can produce all of the tissues of the body, and research with these extraordinarily flexible organic building blocks has deepened scientific understanding of the mechanisms of human development and disease and has led to plans to test promising new treatments for debilitating conditions like spinal cord injuries and age-related blindness. It has, however, been hotly contested in many countries due to its use of "leftover" embryos created through IVF to treat infertility, as well as its potential use of human embryos created through therapeutic cloning (also known as cloning-for-biomedical-research).

In December 2008, two laboratories—one in Japan and one in the United States—announced simultaneously that they had found a way to turn adult human skin cells "embryonic." This apparent success of their revolutionary, surprisingly simple methods (which were nearly identical) means that one day the destruction of human embryos for research may be wholly unnecessary. Until then, there are serious technical hurdles to overcome before these genetically engineered stem cells could ever be used in humans, and for now, embryonic stem cell research continues, as does research with *adult stem cells*—the cells that exist in certain organs and tissues of children and adults to replace or repair cells lost through normal aging and illness.

This chapter looks at features of embryonic and adult stem cells that make them powerful tools for medical research; at the recent successes in turning adult cells "embryonic;" and at two promising areas of investigation into cell-based therapies—one using adult stem cells and one using embryonic stem cells.

TYPES OF STEM CELLS

Stem cells differ from other cells of the body in three important ways. First, they are *unspecialized,* meaning that they do not have the tissue- or organ-specific structures that allow other cells of the body to perform specialized tasks. (Examples of specialized cells include oxygen-carrying blood cells, beating heart muscle cells, *insulin*-producing pancreatic cells, and signal-carrying nerve cells.) Second, stem cells are able to renew themselves through cell division for long periods of time, a process known as *proliferation*. Third, under the right conditions, stem cells can give rise to specialized cells of the body, a process called *differentiation*.

Stem cell differentiation is triggered by internal signals (directed by the cell's genes) and external signals (like contact with other cells or chemicals). Much current work in stem cell research is devoted to understanding whether signals that trigger differentiation of one type of stem cell are similar to signals that trigger other types of stem cells, and whether a specific set of signals invariably produces a single type of specialized cell.

With answers to questions like these, a new world of cell-based therapies could open up for patients living with diseases that are currently incurable. Cell-based therapies are a major part of a new area of research known as *regenerative medicine,* which focuses on the use of stem cells, altered genes, and growth factors to build new, healthy cells and tissues. The hope for cell-based therapies is that scientists may one day be able to take tissue-specific human cells grown in the laboratory and

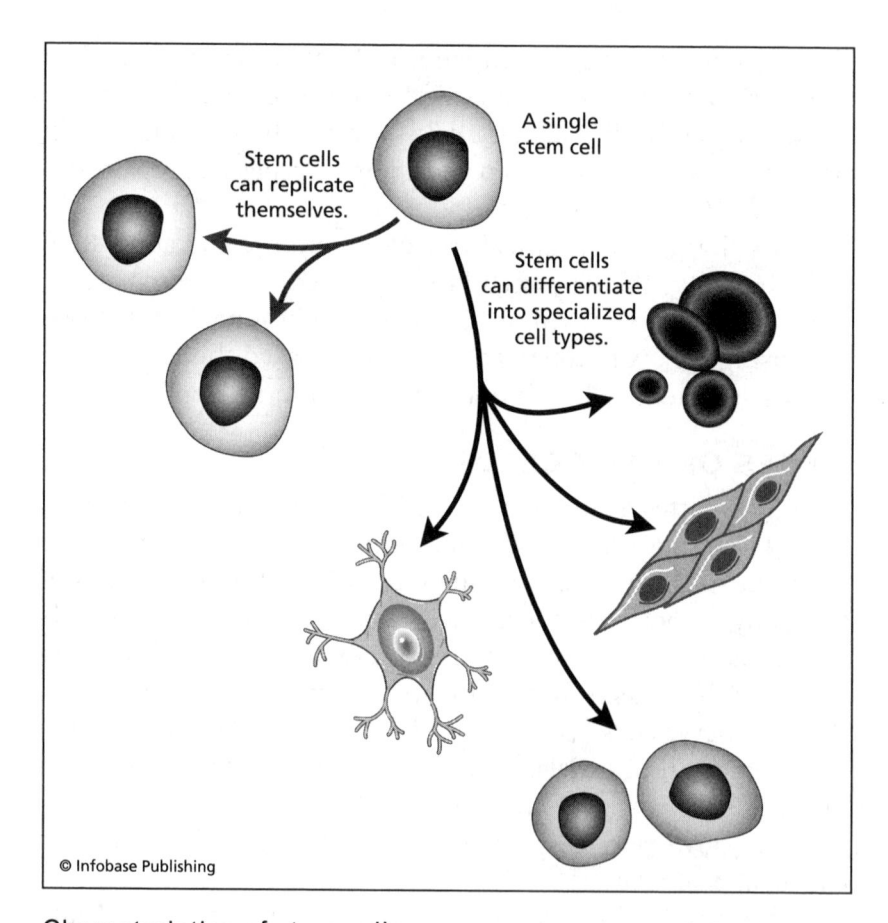

Characteristics of stem cells

introduce them into the human body to replace damaged or diseased ones.

More than two decades worth of laboratory experiments have produced certain protocols, or "recipes," for controlling the laboratory environment in such a way as to produce *directed differentiation* of stem cells into particular cell types. Recipes have been identified, for example, that direct mouse embryonic stem cells to produce *dopamine*-secreting neurons and insulin-secreting pancreatic cells.

Tissue-specific cell cultures may also be used to better understand birth defects or to test the effects of drugs or toxins without subjecting human volunteers or animals to unknown risks. Another promising line of investigation is the possible role of "cancer stem cells" in the growth of tumors, and whether the most successful cancer treatments are ones that specifically target those cells. Cancer stem cells may already have been identified in leukemia, as well as in some cancers producing solid tumors, such as breast and brain cancers.

Embryonic Stem Cells

Stem cells from mouse embryos have been available to researchers for more than 20 years. Not until 1998, however, were human embryonic stem cells successfully isolated and grown in the laboratory. Embryonic stem cells are typically derived from "leftover" embryos that are no longer needed for purposes of in vitro fertilization (IVF). When IVF results in pregnancy, or when hopeful parents undergo unsuccessful procedures and decide not to try again, they face the decision of what to do with their remaining embryos. Some people choose to freeze embryos indefinitely or to discard them, while a very few choose to donate them to "embryo adoption" programs (see chapter 1). Others decide to donate their embryos to research, and they give their informed consent to that effect.

Embryonic stem cells are especially powerful as a research tool because they are *pluripotent*—that is, under the right conditions, they are able to produce all cell types in the body. Embryos used for stem cell research are three- to five-day-old microscopic balls of cells called *blastocysts,* which consist of an outer layer of cells (*trophoblast*), a hollow inside cavity (*blastocoel*), and the *inner cell mass*—a ball of about 30 cells at one end of the cavity, and the source of embryonic stem cells grown in the laboratory.

The inner cell mass is removed and placed in a nutrient soup called the culture medium, which feeds the cells and allows them to reproduce. Until recently, the surface of the laboratory dish was typically treated with mouse embryonic cells—the *feeder layer*—that released nutrients and gave human stem cells a surface on which to attach. But concerns over cross-species transmission of viruses led researchers to find ways of growing cell cultures without using mouse cells. (One argument for

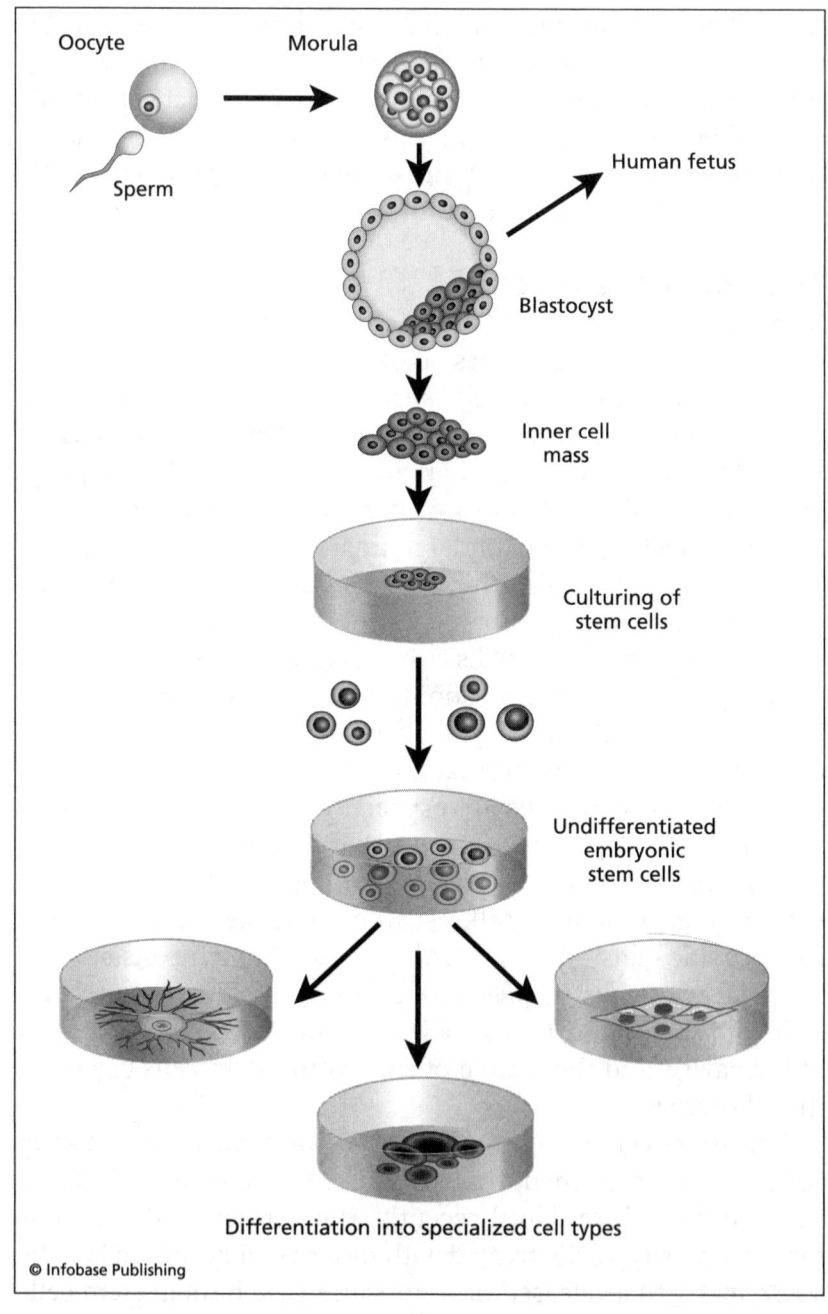

Differentiation of embryonic stem cells into specialized cell types

Stem Cells and Therapeutic Cloning

expanding the list of federally approved stem cell lines is that these older lines contain mouse feeder cells; see section below, "Patchwork Policies on Stem Cell Research.")

After several months of cell division, the original 30 cells of the inner cell mass can produce millions of embryonic stem cells. Once laboratory tests establish that the stem cells are proliferating without differentiating and that they appear to be normal and healthy, they become what is known as an *embryonic stem cell line*.

Embryonic stem cells show great promise for treating certain serious conditions like diabetes and *Parkinson's disease (PD)*, a degenerative condition that affects more than 2

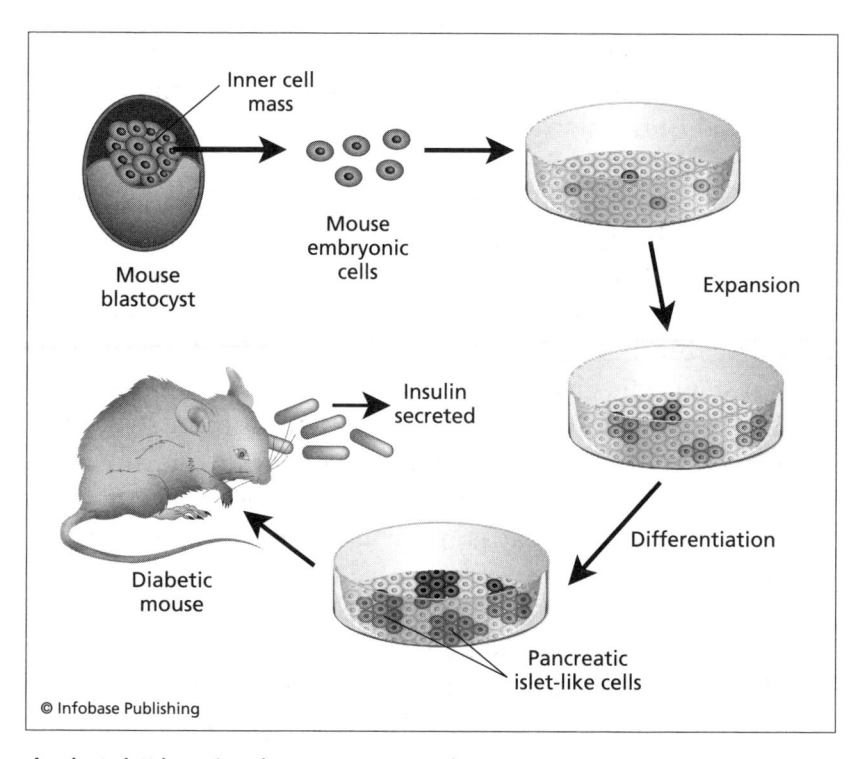

A sketch showing how mouse embryonic stem cells were differentiated into insulin-producing cells and used to treat diabetes in mice

percent of people over the age of 65. In a recent laboratory experiment, researchers triggered mouse embryonic stem cells to differentiate into dopamine-producing neurons—like the nerve cells that are progressively destroyed in PD. The neurons were transplanted into rats suffering from a "rat model" of PD and began producing dopamine, thus significantly alleviating the mice's Parkinson's-like symptoms. In February 2008, scientists were able to turn human embryonic stem cells into insulin-producing cells that controlled blood sugar in diabetic mice.

Adult Stem Cells

Adult stem cells—also known as *somatic stem cells*—have been identified in many different organs and tissues of the body, including the bone marrow, brain, blood vessels, skeletal muscle, liver, and skin. Stem cells in these tissues remain undifferentiated until they are activated to replace specialized cells lost to disease and to normal wear and tear.

Adult stem cells typically give rise to cells of the tissues or organs in which they are located. Under normal conditions in a living animal, adult stem cells follow particular *differentiation pathways*. *Hematopoietic stem cells,* for example, are found in the bone marrow and generate the various kinds of blood cells. Neural stem cells give rise to the major cell types in the brain, and skin stem cells generate the cells that form the protective layer on the surface of the skin. Recent research, however, indicates that some adult stem cell types are more plastic than previously thought—that is, they are able to form cells that make up other tissue types (see table).

The potential for adult hematopoietic stem cells to treat heart disease has generated a great deal of excitement in the past few years. *Coronary artery disease* is caused by blockage of arteries and smaller vessels in the heart by clots and plaque (sticky deposits made up of cholesterol and fatty substances). When these vessels become blocked or narrowed, the heart muscle is deprived of oxygen-carrying blood and can be badly

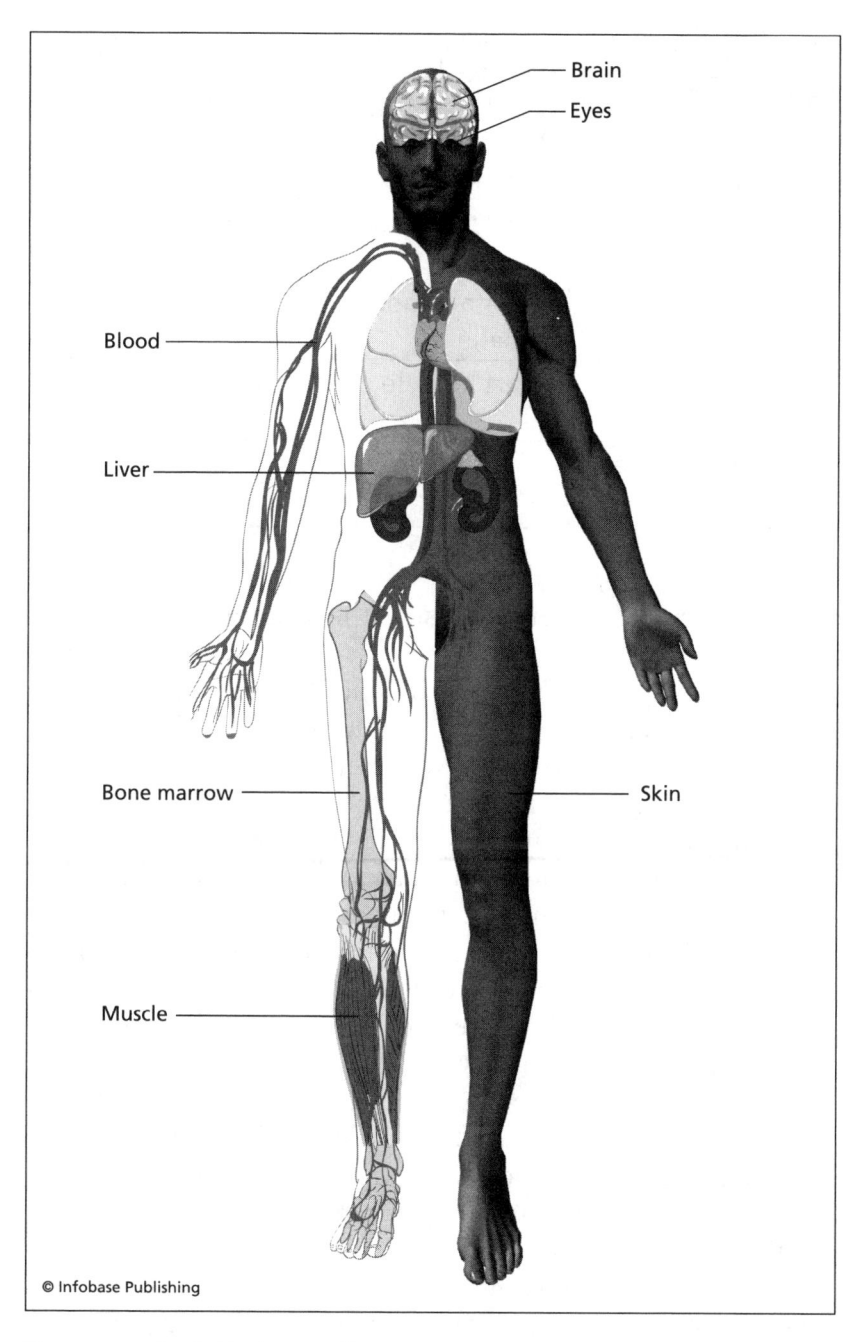

Sources of adult stem cells

ADULT STEM CELL TYPES THAT MAY BE PLURIPOTENT		
Hematopoietic stem cells	**Bone marrow stromal cells**	**Brain stem cells**
may differentiate into:	*may differentiate into:*	*may differentiate into:*
Brain cells	Skeletal muscle cells	Blood cells
Skeletal muscle cells	Heart muscle cells	Skeletal muscle cells
Heart muscle cells		
Liver cells		

injured. Coronary artery disease is the leading cause of death in the United States

Chronic myocardial ischemia (CMI), one of the most severe forms of the disease, strikes between 125,000 and 250,000 Americans every year. In CMI, the tiny vessels that normally distribute blood throughout the heart muscle become constricted, starving the heart of necessary oxygen. This oxygen deprivation can result in a series of small heart attacks that, over time, can cause severe, irreversible damage to the heart.

Steve Myrah was 68 when he signed up for an experiment to test whether adult blood stem cells—filtered from his own blood stream and injected into areas of his heart with poor blood flow—might alleviate his severe CMI. The hope was that the stem cells might promote at least one of three possible improvements for Mr. Myrah and other patients: the growth of new capillaries (the smallest vessels) in the heart; the growth of new arteries and arterioles (tiny arteries); and/or the enlargement of existing arteries and arterioles.

"I'd settle for half as much chest pain as I have now," Mr. Myrah told *UW Health,* the publication of the University of Wisconsin Hospital where he took part in the experiment. He started having chest pains in the 1970s and underwent several heart surgeries over the years, but none of the procedures provided any long-term relief.

Stem Cells and Therapeutic Cloning

In the stem cell experiment, Mr. Myrah and other subjects were injected with a protein that stimulates the release of blood-forming (hematopoietic) adult stem cells, known as *CD34+ cells*, from the bone marrow into the bloodstream. The patients then

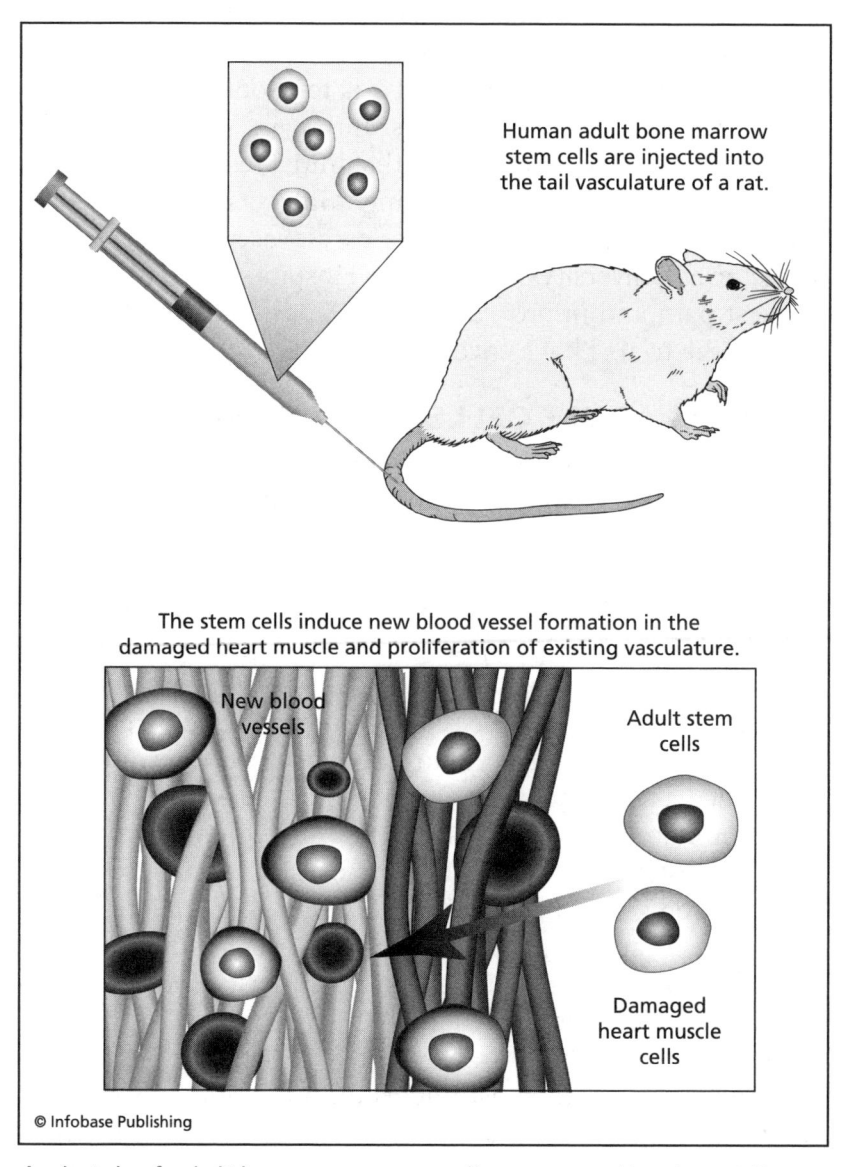

A sketch of adult human marrow cells regenerating heart tissue in mice. The potential of this approach to cure severe forms of heart disease is now being studied in human patients.

underwent a procedure known as *apheresis,* in which a cell separation system filtered CD34+ and other cells from their bloodstream, and laboratory technicians extracted only the CD34+ cells from the mix. The stem cells—or a placebo (fake treatment) for purposes of comparison—were delivered using a special investigational catheter system to areas of the heart muscle suffering from insufficient blood flow.

Experiments are ongoing, and it is too soon to tell whether the procedure will prove successful, but in March 2007, the Wisconsin investigators reported that initial results of the Phase 1 clinical trial had been encouraging. "Subjects reported feeling better," said Amish Raval, head of cardiovascular regenerative medicine at University of Wisconsin Hospital, "with reductions in chest pain and improved exercise capacity during the early stage of the trial. That's encouraging to us."

TURNING ADULT CELLS "EMBRYONIC"

November 2007 brought the breakthrough news that two teams of scientists had found a way to reprogram adult human skin cells back to an apparently embryonic state, and the implications for the future of stem cell research are nothing short of transformative. Researchers hope that the new pluripotent cells—dubbed *induced pluripotent stem cells,* or *iPS cells*—will overcome limitations of both embryonic and adult stem cells. If a patient's own cells could be reprogrammed to produce healthy cells or organs to replace injured or diseased ones, not only could such a technique circumvent ethical questions about the use of human embryos, but it might also prevent the patient's immune system from rejecting new cells or organs, since the patient's body would recognize the cells as the patient's own. Moreover, if large numbers of iPS cells were produced, they might be more efficient candidates for cell-based therapies than adult stem cells, which are found in relatively small numbers in most tissues of the body and are not as versatile.

Perhaps one of the most important implications of the discovery, say researchers, is that the technique will allow them to study the development of cells from patients with complex and debilitating diseases. Scientists could use a skin cell from a person with Alzheimer's to generate neurons, a person with diabetes to generate pancreatic cells, or a person with heart disease to generate heart cells. Seeing abnormal cells in action might provide a wealth of new information about deadly illnesses and how best to treat them. "You cannot really go to a patient and say, 'I want to study your brain,'" Dr. Lorenz Studer, a researcher on neural stem cells at Memorial Sloan-Kettering Cancer Center told the *New York Times* after the breakthrough. "For the first time it gets us access to these cells."

Before the iPS method was hit upon, most researchers believed that the only way to obtain such patient-specific cell lines would be to clone an embryo using the patient's genetic

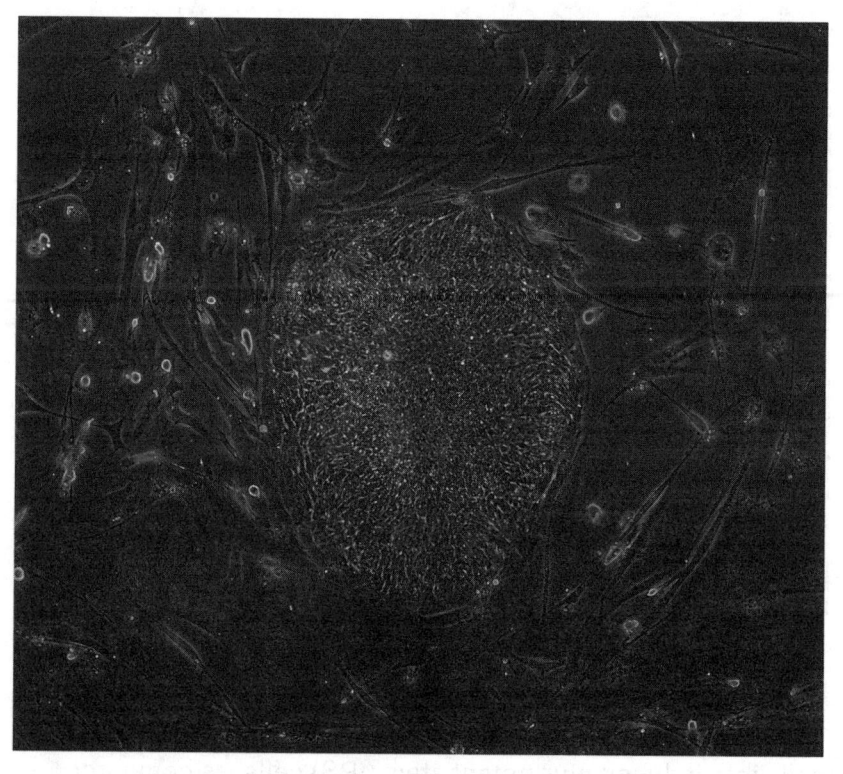

Embryonic-like stem cells derived from skin cells *(Junying Yu/ University of Wisconsin-Madison)*

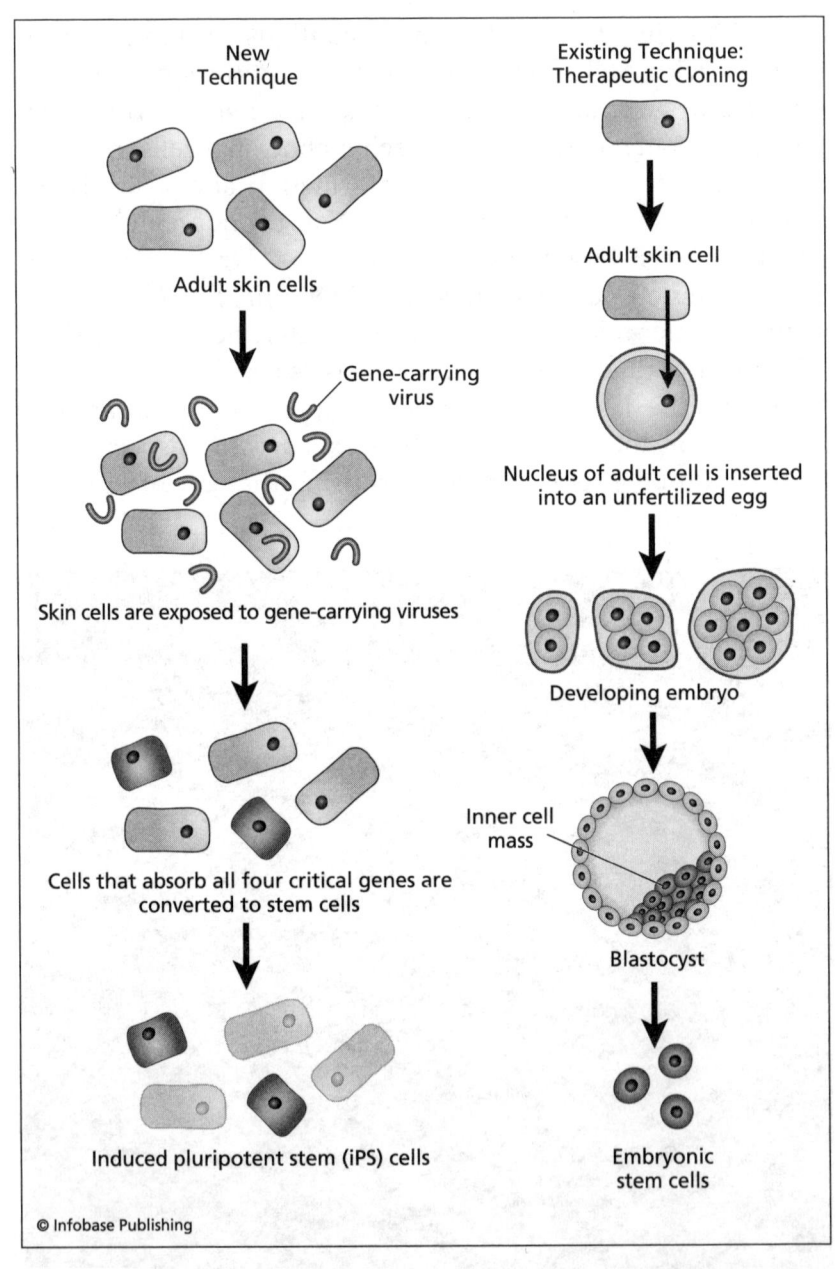

The revolutionary technique that genetically reprograms skin cells into induced pluripotent stem (iPS) cells, as compared to the existing technique—therapeutic cloning by somatic cell nuclear transfer (SCNT)

material and then extract stem cells (see accompanying figure for a step-by-step description of this process). *Somatic cell nuclear transfer,* or *SCNT*—which could be used for reproductive or therapeutic cloning—has already been used to clone animals (see chapter 7 for a discussion of reproductive cloning). A handful of labs around the world have been working on SCNT methods in human cells, but as yet no human stem cell lines have been generated using this process.

The procedure perfected by both teams—one led by Dr. Shinya Yamanaka of Kyoto University and the other led by Dr. James Thomson of the University of Wisconsin—used four genes to reprogram adult skin cells back to a pluripotent state. Each lab started by looking for genes used by embryonic cells but not by adult cells, and they came up with more than 1,000 possibilities. They tested them one by one and identified four that were apparently able to turn skin cells embryonic. "By any means we test them," Dr. Thomson told the *New York Times*, "they are the same as embryonic stem cells."

Thomson and Yamanaka emphasize the need for more testing to confirm that the reprogrammed skin cells are exactly the same as stem cells obtained from embryos. Senator Arlen Specter of Pennsylvania, one of the champions of the stem cell bill vetoed by Mr. Bush in July 2006, agrees. "You've got a life-and-death situation here," Mr. Specter told the *Times*, "and if we can find something which is certifiably equivalent to embryonic stem cells, fine. But we are not there yet."

Even if iPS cells do prove to be indistinguishable from embryonic stem cells, major technical and ethical hurdles will prevent the immediate use of genetically engineered cells in humans. The current technique relies on viral vectors to introduce the genes, and viruses insert themselves into DNA in random locations, thus creating the potential for cancers and other harmful mutations to occur (see chapter 4 for a closer look at viral vectors). One of the four genes used in Dr. Yamanaka's experiments is a known cancer-causing gene. In his preliminary experiments with mice, two of the four genes caused cancer, and 20 percent of the mice died from forms of the disease.

Researchers are hopeful that they will find ways to overcome these limitations. "From the point of view of moving biomedicine and regenerative medicine faster, this is about as big a deal as you could imagine," a leading stem cell biologist at Stanford, Irving Weissman, told the *New York Times* in June 2007. And David Scadden with Harvard Medical School said that the discovery of such a simple technique to reprogram cells "is truly extraordinary and frankly something most assumed would take a decade to work out." Dr. Yamanaka said, "We did work very hard. But we were very surprised."

Since the news that the technique works in skin cells, many researchers have turned their attention to iPS. Dr. Ian Wilmut, the scientist who created Dolly the sheep (see chapter 7 for more on Dolly, the first surviving mammalian clone) announced that he would abandon research on therapeutic cloning and begin working on iPS cells. Wilmut and his team are studying *amyotrophic lateral sclerosis*—a debilitating disease in which progressive loss of nerve cells in the spinal cord and brain cause muscle paralysis—and they hope to use iPS cells to observe the disease in progress and to develop treatments. "All you have to do," Wilmut told *Scientific American* in July 2008, "is take some skin cells from somebody who apparently has inherited the disease, scatter some 'magic dust' on them and wait for three weeks. And you've got pluripotent cells."

Wilmut and others caution that it will be years, in some cases decades, before treatments are developed and approved for use in humans. He likens the future of cell-based therapies to the history of antibiotics and vaccines. "Over a very long period, treatments develop. And I think we should expect the same thing to apply to stem cell-devised treatments, that some will come through in the next few years, but 50 and 100 years from now, people will still be developing new therapies."

PATCHWORK POLICIES ON STEM CELL RESEARCH

Even if the technical and safety issues with iPS cells can be resolved, research with embryos will continue for quite some time. Investigators all over the world have been using embryonic stem cells to research human development, diseases, and potential treatments, and this work will not stop while the new method is perfected. Dr. Thomson told *Nature Reports Stem Cells* in August 2008 that while his focus has shifted to iPS cell lines, the University of Wisconsin will continue to use embryonic stem cell lines since they are the "gold standard" of stem cell research.

"Human ES [embryonic stem] cells created this remarkable controversy," said Thomson, "and iPS cells, while it's not completely over, are sort of the beginning of the end for that controversy. Having a hand in both is very satisfying. One of the legacies is if those culture conditions hadn't been worked out for human embryonic stem cells, iPS cells wouldn't have worked."

Dr. Wilmut points out that to perfect iPS methods, embryos will be needed for some years to come. "The first thing you'll have to do," he told *Scientific American*, "is to look at the cells for the usual quality control things to see that they're expressing the right markers. And for quite a number of years, until you get confidence in the procedure, you'll have to at least form embryo bodies and differentiate them into different lineages. You'll then have to do quality controls to be confident that you've got what you want." He also noted that embryonic stem cells are the only cells that can answer certain questions about fertility problems and early human development.

In June 2007, President George W. Bush issued his second veto of a bill that would have lifted restrictions on federally funded stem cell research. "Destroying human life in the hopes of saving human life is not ethical," he said when he blocked the measure. Even with substantial Republican support for the proposal, the House was unable to achieve the necessary two-thirds vote to override the veto and send the bill back to the Senate. (The House vote was 235-193 in favor of overriding.)

Meanwhile, researchers argued that the 71 federally approved embryonic stem cell lines were not viable clinical research tools. "The problem with federally approved human embryonic stem cell lines is that they contain contaminants from mouse cells such as viruses and mouse proteins," explained Dr.

(continues on page 100)

CELEBRITIES SPEAK OUT ON BOTH SIDES OF THE DEBATE

Celebrity advocates have raised tens of millions of dollars for stem cell research and have campaigned for pro-research candidates in heated races all over the country. Actor Michael J. Fox is a famous example; after being diagnosed with Parkinson's disease in 1991, he stopped acting full time and established the Michael J. Fox Foundation for Parkinson's Research. A high-profile supporter of Senator John Kerry's 2004 presidential race, Mr. Fox has done much to raise awareness of stem cell issues.

Other well-known celebrity supporters of stem cell research have included the late Christopher Reeve, Harrison Ford, Dustin Hoffman, and former first lady Nancy Reagan. "I'm determined to do whatever I can to save other families from this pain," Mrs. Reagan said in 2004 of her husband Ronald's long battle with Alzheimer's disease, and her hope that stem cell research would lead to breakthroughs in Alzheimer's treatment.

Critics of stem cell research maintain that it requires the destruction of a human embryo and therefore is morally unacceptable. Pope John Paul II said in 2000, "A free and virtuous society, which America aspires to be, must reject practices that devalue and violate human life at the very stage of conception until they are dead."

Research advocates counter that blastocysts—the tiny balls of cells left over from IVF procedures—would be destroyed anyway and should be used to research cures for painful and deadly diseases. "What this research has more to do with is not when life begins but when life ends," Dustin Hoffman said at a fund-raiser for diabetes research in 2004. "This research may one day eliminate these diseases from ending people's lives prematurely."

Stem Cells and Therapeutic Cloning

The high-profile debate over stem cell research reached fever pitch in the most unlikely of forums: Game 4 of the 2006 World Series. The series between the Saint Louis Cardinals and the Detroit Tigers became what ABC News called a political "celebrity death match" with the airing of a commercial featuring the Cardinal's starting pitcher Jeff Suppan. "Amendment 2 claims it bans human cloning," Suppan said, "but in the 2,000 words you don't read, it makes cloning a constitutional right. Don't be deceived."

Suppan was referring to a proposed amendment to the Missouri state constitution known as the Stem Cell Research and Cures Amendment (or simply Amendment 2). While the amendment did provide constitutional protections for cloning embryos for biomedical research in the "2,000 words" to which Suppan referred, it also banned cloning for the purpose of creating a child (see chapter 7 for an in-depth treatment of reproductive cloning).

Suppan was joined by other sports and entertainment celebrities, including Arizona Cardinals quarterback Kurt Warner and actors Patricia Heaton of TVs *Everybody Loves Raymond* and Jim Caviezel, who portrayed Jesus in the controversial movie *The Passion of the Christ*.

The advertisement, which cost in the neighborhood of $150,000, countered another that ran during Game 1—a 30-second spot featuring Michael J. Fox, in which he lent his support to Democrat Claire McCaskill's run for U.S. Senate. McCaskill was an outspoken advocate of Amendment 2 (and stem cell research in general), while incumbent opponent Jim Talent was opposed to the constitutional measure.

"They say all politics is local, but it's not always the case," Fox said in the advertisement. "What you do in Missouri matters to millions of Americans—Americans like me."

On November 7, Claire McCaskill won the Senate seat, and Missouri citizens voted to approve Amendment 2.

(continued from page 97)

Eva Zsigmond, associate director of the Laboratory for Developmental Biology at the Institute of Molecular Medicine for the Prevention of Human Diseases (IMM) at the University of Texas Health Science Center. "Mouse contaminants make the federally approved [human embryonic stem cells] unsuitable for human use," Zsigmond said.

Another important problem with the federally approved lines, researchers say, is the age of the stem cell lines. "The current approved stem cell lines are all too old for clinical use," Dr. Rick Wetsel, professor at the Research Center for Immunology and Autoimmune Diseases and director of the Laboratory for Developmental Biology at the IMM, told *HealthLeader,* UT's online health magazine. The older the cell line, the higher the likelihood that genetic mutation has occurred. "You want pure, clean and healthy cells," Wetsel said.

President Bush's veto essentially postponed chances for federal funding into the next presidential term, but many states found ways to fund their own initiatives. California, New Jersey, and Connecticut are among states providing grants for stem cell research, and similar measures have been proposed in other states (such as Texas and Florida). "The lack of federal leadership leaves a vacuum that states are trying to fill on a very piecemeal basis," Representative Andy Meisner of Michigan told the Pew Research Center's Stateline.org in June 2007. Several universities (including Stanford, the University of California, and the University of Wisconsin) have established privately funded programs to support stem cell research. A new state agency called the California Institute for Regenerative Medicine became the biggest U.S. investor in human embryonic stem cell research when they awarded nearly $45 million in research grants in February 2007, and another $271 million in May 2008. California's stem cell program is slated to spend a total of about $3 billion over a decade.

In January 2009, on the heels of President Barack Obama's inauguration, the FDA approved the world's first human trial with embryonic stem cells (for severe spinal cord injuries). Two months after being sworn into office, President Obama issued an executive order lifting the Bush administration's tight restrictions on research with stem cells. At that ceremony, Obama promised that his administration would "make scientific decisions based on facts, not ideology" and expressed the hope that Congress would pass legislation in support of stem cell research.

EMBRYONIC STEM CELLS: A CURE FOR AGE-RELATED BLINDNESS?

Scientists are working to clear the technical roadblocks to iPS cell treatments in humans, but in the meantime, should embryonic stem cell treatments be made available to patients as soon as they are developed? Countries encouraging research with embryonic stem cells plan to test them in humans soon. Researchers in Britain, for example, hope that within two years they can select the first patients for a promising stem cell treatment for *age-related macular degeneration (AMD)*.

AMD is the leading cause of blindness and impaired vision in people over 60 in the United States and most Western countries. Advanced AMD affects approximately 1.8 million Americans age 40 and older. Another 7.3 million people with intermediate AMD are at substantial risk for loss of vision, and an estimated 2.9 million people will be living with advanced AMD by the year 2020—unless researchers find a cure.

The *macular area* is the central part of the retina (the inner lining of the back of the eye) that is rich in *cones*—cells that detect color and fine detail. *Dry macular degeneration* is the most common type of AMD, and it affects approximately 90 percent of people who suffer from the disease. In dry AMD, the light-sensitive cells of the macular area begin to deteriorate, resulting in a spotty loss of "straight ahead" vision.

The London Project to Cure AMD is a collaborative effort of scientists all over Britain, including doctors at London's Moorfields eye hospital who have already restored vision using healthy cells harvested from patients' own eyes. That procedure is difficult and can involve only a small number of cells, but for some patients the results have been "spectacular" according to Dr. Lyndon da Cruz at Moorfields. People "have got their driving license back" and "have gone on to reading newspapers. That gives some taste of what a perfect transplant might do."

Researchers hope that the "perfect transplant" will be achieved through the differentiation of embryonic stem cells into retinal cells grown in small transplant "patches," which would remove the need for complicated harvesting procedures from patients' eyes. Experiments with embryonic stem cells have proven highly successful in rats, and da Cruz and others hope that the procedure will be ready to try in humans within two years.

"If it hasn't become routine in about 10 years it would mean we haven't succeeded," Dr. da Cruz told reporters at the study's launch on June 5, 2007. "It has to be something that's available to large numbers of people."

The study was made possible by an $8 million contribution from an anonymous American donor who was discouraged by tight restrictions on stem cell research in the United States.

SUMMARY

In March 2009, President Obama's executive order lifting restrictions on stem cell research drew direct fire and quick praise from the expected political corners. "The administration now steps onto a very steep, very slippery slope," Douglas Johnson, legislative director for the National Right to Life Committee, said in advance of the president's order. "Many researchers will never be satisfied only with the so-called leftover embryos."

The late senator Edward M. Kennedy spoke on behalf of supporters of stem cell research and patients who stand to benefit. "Today, an extraordinary medical breakthrough was achieved with the stroke of a pen. With today's executive order, President Obama has righted an immense wrong done to the hopes of millions of patients."

Against this volatile political backdrop, new cell-based advances continue to make regular headlines. January 2008 saw the creation of a living, beating rat heart at the University of Minnesota; the heart was constructed from the outer structure and valves of a dead rat heart injected with fresh heart cells from newborn rats. And in August 2008, researchers with the

Harvard Stem Cell Institute reprogrammed ordinary pancreatic cells in mice into insulin-producing cells.

The next chapter looks at the closely related issue of abortion—another ethical controversy stoked by scientific ambiguity about when life begins.

6

Abortion and Emergency Contraception

Of the medical issues considered in this volume, perhaps the most contentious is abortion. On one side of the debate are abortion opponents who believe that human life begins at conception and that it is their duty to protect innocent lives; on the other side are abortion rights advocates who believe that the health and welfare of pregnant women must take precedence. Most Americans express opinions somewhere between the two hard-line positions, neither defending nor condemning abortion in all cases.

This chapter begins with one couple's decision that became national news a decade before abortion was legalized in the United States, and it continues with a survey of philosophical and medical positions on abortion. The chapter ends with two defining legal and political battles that bookend the last 35 years of debate—the Supreme Court decision on *Roe v. Wade,* and the recent federal ban on so-called *partial-birth abortion.*

BEFORE ABORTION WAS LEGAL: THE SHERRI FINKBINE STORY

In 1962 Sherri Finkbine, mother of four healthy children, was pregnant for a fifth time. She had intended to carry the baby to term before she learned something frightening: Europe had seen a dramatic increase in babies born with severe deformities, and the tragedy had been linked to a common tranquilizer which, prior to these heartbreaking outcomes, had been assumed to be perfectly safe. The problem ingredient was *thalidomide,* and it caused babies to be born with malformed or absent limbs, blindness and deafness, and sometimes seriously impaired internal organs.

Ms. Finkbine learned that her husband's tranquilizer, which she had been taking to help her sleep, contained thalidomide. Her doctor told her, "The odds are so against you that I am recommending termination of pregnancy." This was before the Supreme Court decision on *Roe v. Wade* in 1973 that effectively legalized abortion in the United States. In most states, including Arizona where Ms. Finkbine lived, abortion was illegal unless the life of the mother was in danger, but Ms. Finkbine's doctor felt that she would have little problem receiving approval for an abortion from the small medical board at the Phoenix hospital.

Ms. Finkbine, a local TV personality, believed that she should inform other American women

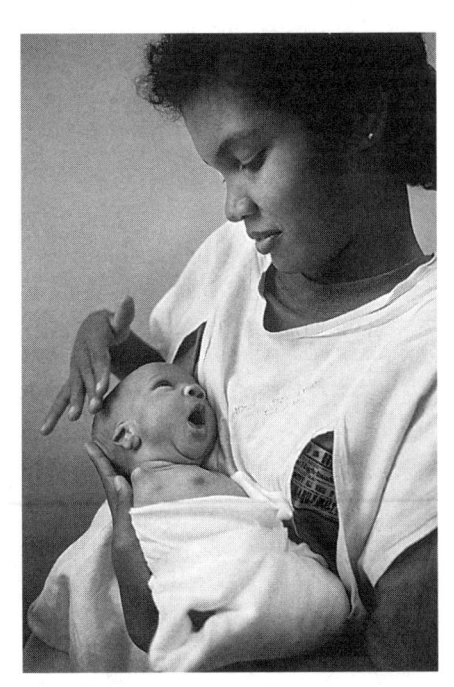

Brazilian Luciene Da Dores with her son, Rafael, who suffers the effects of thalidomide. Though the drug is known to cause severe birth defects, it is used to treat leprosy and was approved in 2006 by the FDA to treat multiple myeloma. *(John Maier, Jr./The Image Works)*

about the risks of thalidomide, but she wished to remain anonymous. She contacted a local newspaper, which ran her story under the headline, "Baby-Deforming Drug May Cost Woman Her Child Here." The paper did not use her name, but the story was picked up nationally and soon her identity was exposed. The medical board withdrew their approval for the abortion since they did not think their decision would be upheld in court.

Over the coming weeks, Ms. Finkbine became a lightning rod for intensely angry sentiments from abortion opponents. The official Vatican newspaper, *Il Osservatore Romano,* called her and her husband murderers. The family received many abusive letters including one that said, "I hope someone takes the other four children and strangles them, because it is all the same thing."

Eventually the Finkbines flew to Sweden for the procedure. The attending surgeon told Ms. Finkbine that the fetus was severely deformed and would not have survived. According to the doctor's report, the baby had no legs and only one arm.

PERSPECTIVES ON ABORTION

Throughout the history of Western civilization, a number of different religious and medical theories have dominated the debate about when human life begins. Early Judeo-Christian writings appear to take the point of formation (i.e., when the fetus begins to *look* human) as the point at which the fetus attains full moral status, while ancient English law recognized both the formation of the fetus and its quickening—when movements are first felt by the pregnant woman—as crucial. Henry de Bracton, a famous 13th-century juridical writer, described the importance of both milestones in his *On the Laws and Customs of England*: "If one strikes a pregnant woman or gives her poison in order to procure an abortion, if the foetus is already formed or quickened, especially if it is quickened, he commits homicide."

Five hundred years later, British legal scholar William Blackstone described the importance of quickening to the historical development of abortion laws: "Life . . . begins in contemplation of law as soon as an infant is able to stir in the mother's womb. For if a woman is *quick* with child, and by a potion, or otherwise, killeth it in her womb; or if any one beat her, whereby the child dieth in her body, and she is delivered of a dead child; this, though not murder, was by the ancient law homicide or manslaughter. But at present it is not looked upon in quite so atrocious a light, though it remains a very heinous misdemeanor."

The Current Roman Catholic Position

It was not until 1869 that Pope Pius IX overturned centuries of Catholic dogma, inspired by Thomas Aquinas and before him, Aristotle, that human "ensoulment" does not occur until the fetus attains an intellectual/rational soul (after first possessing a vegetative/nutritive soul and then a sensitive soul, which is also possessed by many animals). It was Pius IX who established the current official Vatican position that life begins at conception.

Medical ethicists discuss two problems for this view: first, the existence of twins, whose genetic identities do not separate until after conception, and second, the existence of *genetic mosaics,* also called "chimeras." A genetic mosaic is an organism in which different cells of the body are genetically different. This can happen if two embryos fuse or if a mutation occurs when chromosomes are in the process of dividing.

John Finnis, a Catholic medical ethicist, argues that *most* persons (he defines them as "living human individuals") begin at conception, and that twinning and mosaics are unusual cases that are not problematic for the case against abortion. Biologically, he says, "one always finds just individuals. If these split, or combine to form a mosaic, one then simply finds one or more different individuals. Twinning is an unusual way of being generated . . . Being absorbed into a mosaic would presumably be an unusual way of dying."

Twins and mosaics—as well as evidence that the "moment" of conception actually takes several hours—have generated debate within the church, but Gregory Robbins, historian of religion at the University of Denver, told *Nature Magazine* in April 2005 that a softening of the Vatican's position is unlikely anytime soon. "Ensoulment," he said, "including the provision of souls for what will become twins or triplets, presumably takes place within the purview of God's foreknowledge. That does not seem to pose much of a theological problem. Embryonic and therapeutic stem-cell research do." Abortions do as well, at any stage of development, including abortions intended to protect the health of the mother. (Measures to protect her may be taken even if they result in the death of the fetus, so long as the death of the fetus is never intended.)

When Does Personhood Begin?

Most Americans—including many practicing Catholics—take a more middle-of-the road approach to abortion than does the Vatican. A July 2007 ABC News/Washington Post poll asked adults nationwide, "Do you think abortion should be legal in

---

HUMAN-ANIMAL HYBRIDS AND THE RIGHT TO LIFE

In June 2007, the official Roman Catholic position that life begins at conception took the right-to-life debate in a rather unexpected direction, when the Catholic bishops of England and Wales declared that human-animal hybrid embryos should be regarded as human life, and that if the biological mothers of these embryos want to carry them to term, they should be allowed to do so.

At issue was draft legislation that would grant scientists permission to create so-called chimeras, named for the mythical creature that was part lion, part goat, and part serpent. The proposed law prohibits the creation of "true hybrids"—embryos created by combining eggs and sperm from humans and animals—but permits the creation of *cytoplasmic hybrid embryos* by transferring human DNA into animal eggs that have been stripped of most of their genetic information. It also allows for animal DNA to be introduced into human embryos and for human embryos to be mixed with one or more animal

all cases, legal in most cases, illegal in most cases, or illegal in all cases?" The majority of Americans (62 percent) took one of the softer positions—that is, that they believed abortion should either be "legal in most cases" (34 percent) or "illegal in most cases" (28 percent). Only 37 percent took a hard line either way, answering either "legal in all cases" (23 percent) or "illegal in all cases" (14 percent). And in May 2007, a Gallup poll showed that 49 percent of Americans considered themselves "pro-choice," while 45 percent considered themselves "pro-life."

Results of polls on abortion fluctuate somewhat from month to month (see table), reflecting the tenor of political debates and the fundamental ambivalence many people feel toward the issue.

---

cells but makes it illegal to grow a hybrid embryo longer than two weeks or to implant one in a woman's womb.

The bishops, who are opposed to the creation of any embryo for research in the first place, also wish to prevent the destruction of embryos after they have been created. The bishops submitted their opinion to a parliamentary committee reviewing the draft legislation. "At the very least," the bishops wrote to the committee, "embryos with a preponderance of human genes should be assumed to be embryonic human beings, and should be treated accordingly. In particular, it should not be a crime to transfer them, or other human embryos, to the body of the woman providing the ovum, in cases where a human ovum has been used to create them. Such a woman is the genetic mother, or partial mother, of the embryo; should she have a change of heart and wish to carry her child to term, she should not be prevented from doing so."

During committee debates on the proposed legislation, Tory MP Edward Leigh led a fight to ban hybrid embryos, calling them "a step too far." The ban was defeated in the House of Commons in May 2008.

ABC NEWS/WASHINGTON POST POLL ON ABORTION, JULY 18–21, 2007

"Do you think abortion should be legal in all cases, legal in most cases, illegal in most cases, or illegal in all cases?"

	Legal in All Cases %	Legal in Most Cases %	Illegal in Most Cases %	Illegal in All Cases %	Unsure %
7/07	23	34	28	14	2
2/07	16	39	31	12	2
12/05	17	40	27	13	3
4/05	20	36	27	14	3
12/04	21	34	25	17	3
5/04	23	31	23	20	2
1/03	23	34	25	17	2
8/01	22	27	28	20	3
6/01	22	31	23	20	4
1/01	21	38	25	14	1
9/00	20	35	25	16	3
7/00	20	33	6	17	4
9/99	20	37	26	15	2
3/99	21	34	27	15	3
7/98	19	35	29	13	4
8/96	22	34	27	14	3

Source: PollingReport.com

When considering the abortion issue, a central question is: When does personhood begin? The previous section looked at the current official Catholic position that life begins at conception, and therefore that any abortion is morally wrong—that it is the purposeful taking of a human life, and that there is no case in which it would be considered acceptable.

For many the line is not so clear. Is an embryo or fetus a person because it possesses a unique set of DNA that provides the biochemical blueprint for the development of a unique individual?

Or is a fetus not a separate individual until it can live outside the womb (until it is viable), a line that shifts continually as neonatal intensive care technology improves? Or does personhood appear at some point in between—perhaps with the ability to consciously perceive sensory input, or the ability to experience pain, or some combination of biological and/or cognitive factors? If so, it would be difficult—probably impossible—to identify a precise moment in each pregnancy when the fetus becomes a person.

Though much political debate has centered on late-term abortions, the vast majority of abortions occur in early pregnancy. The CDC reports that 88 percent of abortions in 2005 were performed before 13 weeks (in the first trimester), with only 3.7 percent occurring at 16 to 20 weeks and only 1.3 percent after 21 weeks (in the second half of pregnancy).

Many feminists see the abortion controversy as a piece of the larger struggle for the liberation of women from male dominance, arguing that control over reproductive health and childbearing is central to control over all other aspects of a woman's life. Since birth control is not perfect, restrictions on abortion mean that women will never enjoy as much control over their lives as men do. Though the fetus may have some moral standing, its significance cannot be separated from its relationship to the pregnant mother.

In her 1991 article, "Abortion Through a Feminist Ethic Lens," Susan Sherwin argues that fetuses "are not individuals housed in generic female wombs, nor are they full persons at risk only because they are small and subject to the whims of women. Their very existence is relational, developing as they do within particular women's bodies, and their principal relationship is to the women who carry them."

The mistake most commentators make, Sherwin argues, is to assume that there is some particular feature—genetic heritage or self-awareness, for example—"by which we can neatly divide the world into the dichotomy of moral persons (who are to be valued and protected) and others (who are not entitled to the same group privileges)," and that such simplistic divisions ignore the importance of social relationships. "Personhood is a social category," she argues, "not an isolated state," and while fetuses are not full persons because they have not begun to develop social relationships outside the pregnant woman, newborns "are immediately subject to social relationships, for they are capable of communication and response in interaction with a variety of other persons."

Some abortion rights advocates argue that personhood actually begins well after birth, since it requires not just awareness of sensations or the ability to perceive pain, but characteristics like self-awareness and the ability to solve problems, and therefore women—who do have full moral status as persons—should have the right to abort at any stage of a pregnancy to protect their own interests and well-being. This position is somewhat stronger than many Americans take, since it excludes some young children and mentally disabled adults from the category of "persons" and does not recognize newborn babies as having rights.

Some theorists argue that the core issue is not personhood, but an individual's right to a valuable future. Don Marquis, a philosopher at the University of Kansas, offered an influential form of this argument in his 1989 article, "Why Abortion Is Immoral," published in the *Journal of Philosophy*. A fetus can be assumed to have a future of value just like an adult human, Marquis argued, and therefore abortion "is presumptively very seriously wrong, where the presumption is very strong—as strong as the presumption that killing another adult human being is wrong." Marquis does not take issue with contraception, since before conception there is no individual to be robbed of his or her future.

Marquis's argument, however, would not satisfy many feminist theorists since it does not give primacy to social relationships and a person's ability to form them. They would respond that a fetus cannot be considered a separate individual because of its intimate relationship with the pregnant woman.

EMERGENCY CONTRACEPTION

After several years of debate, in August 2006 the FDA approved over-the-counter sales of the so-called *morning-after pill* for women over 18. The pill, *Plan B,* had already been available by prescription for several years, but women's health advocates argued that women needed unrestricted access to the drug for it to be effective in preventing unwanted pregnancies.

"Because taking emergency contraception is time sensitive, it is important for women to be able to have access to it 24 hours a day seven days a week," Gloria Feldt, president of the Planned Parenthood Federation of America, told the *New York Times* in December 2003. "If every woman of reproductive age had access to it when she needs it, we could prevent half the unintended pregnancies and half the abortions."

Plan B can reduce the risk of pregnancy by 89 to 95 percent after unprotected sex, but its effectiveness declines the longer a woman waits to take it. The sooner she takes it after unprotected sex, the more effective it is. "Emergency contraceptives don't work if, like condoms, they're left in the drawer," noted James Trussell, director of Princeton University's Office of Population Research.

Plan B consists of two pills made from a synthetic hormone, levonorgestrel, which has been used in birth control pills for more than 35 years. To be effective, the first pill should be taken as soon as possible within 72 hours of unprotected sex, and the second pill 12 hours after the first. The pills usually prevent pregnancy by preventing ovulation (release of an egg from the ovary) or by preventing fertilization of the egg. In rare cases, it may stop an already fertilized egg from implanting in the uterine lining, which many opponents of the drug consider to be an early form of abortion.

"When it comes to contraception as a policy issue—access, availability—the Catholic bishops do not get involved in that debate," said Cathy Cleaver Ruse, a spokeswoman for the bishops. "But when it comes to abortion, that's a different matter. It's far greater than just a religious issue. It's a human rights issue."

The FDA's policy since the 1970s has been that pregnancy begins at implantation, since there is no effective way to test for pregnancy before that point. The agency has not approved over-the-counter distribution for women under the age of 17, arguing that adolescents would benefit from seeing a doctor before taking the drug. Barr Pharmaceuticals, the drug's manufacturer, has said that it will continue to study the drug's use in adolescents with the goal of eliminating the age restriction.

WHEN ABORTION BECAME LEGAL: *ROE V. WADE*

The circumstances of Norma McCorvey's young life were cruel. Sexually and emotionally abused as a young child, raped as a teenager, she married an abusive husband at the age of 16 and gave birth to two daughters before the age of 20. The first child was raised by Ms. McCorvey's mother, the second by her second husband. Ms. McCorvey turned to drugs and alcohol at a young age.

By 1970 she was 21, pregnant again, and unmarried with very little money. She sought an abortion, but her home state of Texas considered the procedure a criminal act unless deemed necessary to save the life of the mother—just as Arizona had eight years earlier when the Finkbines sought to terminate their pregnancy. Ms. McCorvey wanted to travel to California for the procedure, but she had no money for the trip.

Two lawyers, Linda Coffee and Sarah Weddington, asked Ms. McCorvey if she wished to join a lawsuit against Dallas district attorney Henry Wade challenging the constitutionality of the Texas law. Ms. McCorvey agreed, and to conceal her identity, she became "Jane Roe" for the purposes of the proceedings. *Roe v. Wade,* one of the most influential cases in American legal history, was underway.

A federal district court ruled that the Texas law was unconstitutional, but District Attorney Wade appealed the decision to the Supreme Court. In the meantime, Ms. McCorvey carried her pregnancy to term and gave birth to another daughter, whom she gave up for adoption.

In January 1973, the Supreme Court handed down a 7-2 decision that the Texas law was indeed unconstitutional. The ruling recognized states' rights to regulate abortion, but it also specified ways that states could not restrict abortion without violating a woman's constitutional right to privacy. Using the standard medical division of pregnancy into trimesters as a guide, the Supreme Court ruled that during the first trimester

Abortion and Emergency Contraception

(approximately 12 weeks) of pregnancy, states cannot restrict a woman's right to an abortion at all. During the second trimester, the Court said, only restrictions to protect the health and safety of pregnant women are permissible, and during the third trimester—when the fetus could be considered viable outside of the woman's body—restrictions on abortion are permissible, but only if those restrictions preserve a woman's health.

Abortion rights advocates welcomed the Supreme Court ruling, which effectively legalized abortion in the United States and elevated women's health to priority status. Abortion

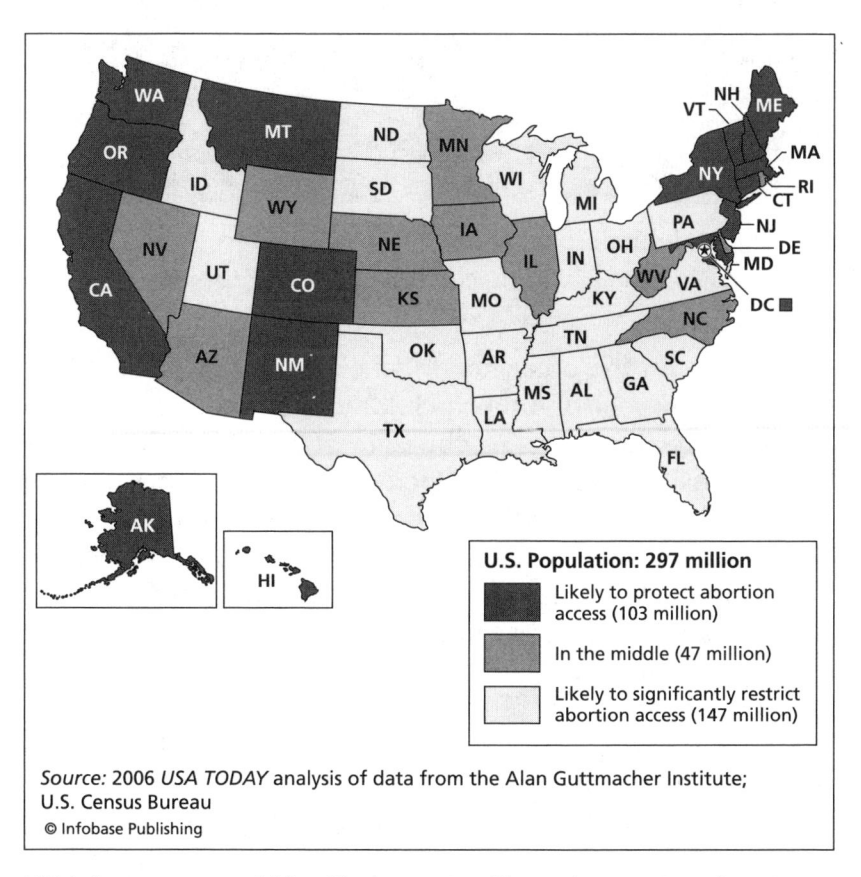

U.S. Population: 297 million
- Likely to protect abortion access (103 million)
- In the middle (47 million)
- Likely to significantly restrict abortion access (147 million)

Source: 2006 *USA TODAY* analysis of data from the Alan Guttmacher Institute; U.S. Census Bureau
© Infobase Publishing

Which states would be likely to significantly restrict abortion access and which would be likely to protect it if *Roe v. Wade* is overturned? *(Sources:* USA Today *analysis of data from the Alan Guttmacher Institute, U.S. Census Bureau)*

Abortion rights defenders and opponents demonstrate outside the U.S. Supreme Court in 2005. *(Manuel Balce Ceneta/AP Images)*

opponents, on the other hand, believed that the ruling sanctioned the murder of unborn children, and they have worked ever since to shorten its reach. Some states now require waiting periods; others require doctors to say certain things to patients as part of the informed consent process. Most recently, a federal ban on a particular abortion procedure—so-called partial-birth abortion—was upheld by the Supreme Court (see the discussion of that case in the next section).

In a twist to the *Roe v. Wade* story, Norma McCorvey experienced a radical change of heart after the Supreme Court decision. She was employed at a women's clinic in the Dallas area when an Operation Rescue group protested the clinic's work and she met Philip Benham, an evangelical preacher who led the local group. She agreed to attend one of his church services, and converted to Christianity on her first visit. Before long she was working for Operation Rescue. Ms. McCorvey has since told the media that she is committed to encouraging women to seek alternatives to abortion.

THE PARTIAL-BIRTH ABORTION BAN UPHELD: *GONZALES V. CARHART*

Out of the approximately 1.3 million U.S. abortions performed in the year 2000, only some 0.17 percent (about 2,200) involved a method known as *intact dilation and extraction* (*IDX* or intact D&X)—or as it is more controversially known in American political debates, partial-birth abortion. With IDX, the brain of the fetus is evacuated before the body is delivered vaginally. This may be done for several reasons. The procedure does not, for example, require that women undergo abdominal surgery or labor, and it produces an essentially intact body over which parents can grieve.

Federal legislation banning the procedure was upheld by the Supreme Court in the spring of 2007 in a 5-4 decision, and though abortion rights advocates say the Court's decision will harm the health of women, abortion opponents argue that it is in the best interests of women's emotional health. "While we find no reliable data to measure the phenomenon, it seems unexceptional to conclude some women come to regret their choice to abort the infant life they once created and sustained," Justice Kennedy wrote, echoing sentiments in a brief filed by the Justice Foundation, a conservative advocacy firm. "Severe depression and loss of esteem can follow."

Women's rights advocates called the ruling condescending to women, and Planned Parenthood called the Justice Foundation brief "extraneous." Justice Ruth Bader Ginsburg wrote in her dissenting opinion that the Court had invoked antiabortion sentiments "for which it concededly has no reliable evidence."

Abortion opponents hope that the Supreme Court decision on IDX will open the door to more state legislation limiting abortion rights. Clarke D. Forsythe, president of Americans United for Life, called the Court's ruling "a green light for enhanced informed consent."

Abortion rights advocates see such state initiatives differently. "Informed consent is really a misleading way to characterize it," Roger Evans, senior director of public policy litigation and law for Planned Parenthood, told the *New York Times* in May 2007. "To me, what we'll see is an increasing attempt to push a state's ideology into a doctor-patient relationship, to force doctors to communicate more and more of the state's viewpoint."

SUMMARY

Now, 35 years after *Roe v. Wade,* most Americans take a centrist position on the issue of abortion, and some conservative politicians consider themselves supporters of abortion rights. (Rudolph Giuliani, former mayor of New York City, is a well-known example.) But in many states, the process of seeking an abortion has become increasingly difficult. South Dakota, for example, passed a law requiring that doctors tell women an abortion will terminate a "whole, separate, unique, living human being."

Reva B. Siegel, a law professor at Yale who has studied this state-by-state legislative effort to restrict abortion, told the *New York Times* in May 2007 that abortion opponents are mixing "the modern language of trauma and women's rights" with "some very traditional ways of understanding women"—referring to the paternalistic view that women who seek abortions do not understand the risks and need to be protected from themselves. Justice Kennedy's language in the *Gonzales v. Carhart* ruling, Siegel said, was "beyond Alice in Wonderland: criminalize abortion to protect women."

Abortion opponents characterize their new approach differently. "We think of ourselves as very pro-woman," said the president of the National Right to Life Committee, Wanda Franz. "We believe that when you help the woman, you help the baby."

The next chapter highlights two extreme scenarios in assisted reproduction—cloning and ectogenesis—and what medical measures might be considered ethically acceptable when the intention is to produce, rather than to prevent, a human life.

Reproductive Cloning and Ectogenesis

When the face of Dolly the sheep—the first cloned mammal—appeared on the front pages of the world's major newspapers and magazines in 1997, the hypothetical debate over reproductive cloning suddenly became real. If sheep could be cloned, what was to stop the cloning of humans? Two distinct reasons for cloning were at issue: reproductive cloning (also known as cloning-to-produce-children), and therapeutic cloning (or cloning-for-biomedical-research).

Chapter 1 looked at the ethical debate surrounding therapeutic vs. reproductive cloning in the context of the "problem of scope" (the moral status of the fetus), and chapter 5 examined therapeutic cloning to produce embryonic stem cells for research purposes. That chapter also looked at groundbreaking research on turning adult skin cells "embryonic," which may one day replace the need for therapeutic cloning altogether. Even if that day comes, human cloning for reproductive purposes is still a real possibility. Cloning technology has already created new animals like Dolly, and some people hope that the techniques will be perfected to such a degree that favorite pets can be cloned on a regular basis.

Dolly the sheep, the first mammal successfully cloned from an adult cell, with her three lambs at the Roslin Institute in Edinburgh, Scotland *(Topham/PA/The Image Works)*

What is to stop the cloning of favorite humans? This chapter explores the brief history of reproductive cloning and its possible futures, along with the controversial topic of *ectogenesis* (the gestation of fetuses in artificial wombs)—the next reproductive revolution on the horizon.

DOLLY: THE FIRST ANIMAL CLONED FROM AN ADULT

Ian Wilmut, the Scottish scientist who cloned Dolly, could not have produced a less threatening poster child for his groundbreaking research than the fuzzy, sweet-faced Finn Dorset lamb, but the overall tone of press coverage was cautionary—even fearful. The world was stunned by Wilmut's discovery at the Roslin Institute in Edinburgh, Scotland. Few legitimate researchers

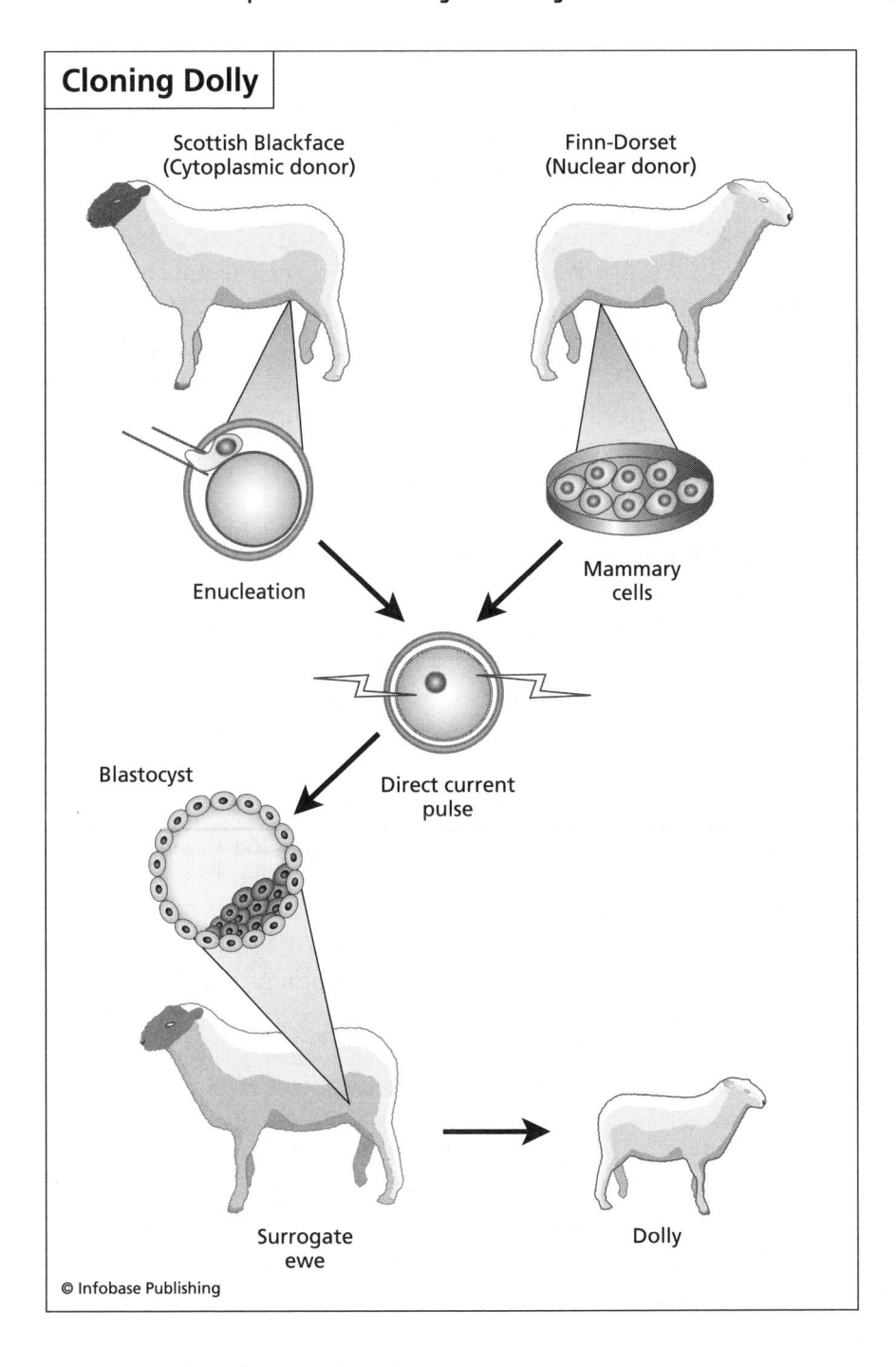

Reproductive Cloning and Ectogenesis 121

Cloning Dolly

Scottish Blackface (Cytoplasmic donor)

Finn-Dorset (Nuclear donor)

Enucleation

Mammary cells

Direct current pulse

Blastocyst

Surrogate ewe

Dolly

© Infobase Publishing

had been trying to clone mammals from adult cells, since most believed that success was unlikely any time soon. Then skeptics were blown away by the undeniably real existence of Dolly, and by the real possibility that humans could be cloned from adult (somatic) cells.

Dolly "is the category of experiment that bends your mind," said Zena Werb, a developmental cell biologist at the University of California, San Francisco in *Science* magazine in March 1997, soon after Dolly's arrival. The frisky little lamb was the genetic twin of an adult Finn Dorset ewe. She had no genetic father and was produced by a process that combined mammary cells from the Finn Dorset and egg cells retrieved from a Scottish Blackface ewe.

Wilmut was able to stop the mammary cells' process of cell division by placing them in a low-nutrient medium, thus essentially programming them back to a more developmentally plastic state. (See chapter 5 for a detailed discussion of stem cell pluripotentcy, or plasticity.) The DNA was removed from the egg cells, and the eggs—now empty of nuclei—were mixed with the mammary cells. A weak electric current fused some of the cells together, and a second electric pulse started the process of cell division. The fused cells developed into early-stage embryos, and Wilmut was able to implant one of them into the womb of a third sheep, also a Scottish Blackface ewe. The pregnancy developed normally, and the surrogate mother gave birth to Dolly—a genetic clone of the Finn Dorset.

Much of the coverage in the scientific literature and popular press focused on her miraculous "conception" and birth, but the lead-up to Dolly was far from clean. From an animal rights perspective, the process had been quite cruel. Several surrogate mothers were killed to perform post-mortem autopsies on fetuses, and one lamb died at birth. Out of 277 attempts to fuse cells to form healthy embryos, Dolly was the lone survivor.

Dolly lived her life at the Roslin Institute, where she gave birth to six lambs of her own. At the age of five, she developed painful arthritis, which was treated successfully with anti-inflammatory drugs. This was not the end of her health troubles, and she was euthanized at the age of six because of a progressive lung disease. There was curiosity at the time about whether she developed such severe health problems because she had been produced by cloning. (A Finn Dorset sheep typically has a life expectancy of about 12 to 15 years.) Upon autopsy, it was discovered that she suffered from a type of lung cancer that is caused by a retrovirus and is fairly common among sheep who live indoors. (Dolly slept inside for security reasons.) Moreover, according to Roslin scientists, other sheep in her flock had died of the same disease.

The opinion of Roslin researchers notwithstanding, some scientists hypothesize that Dolly may have been born with a genetic age of six years—the age of the sheep from which she was cloned. Evidence that supports this idea includes shortened *telomeres* (protective regions of DNA at the ends of chromosomes) on Dolly's DNA, a typical result of the aging process. The Roslin Institute stated that her health was screened regularly and that there was no indication of problems related to advanced aging, but Wilmut himself acknowledged uncertainty about her true age from the start. Asked by the *New York Times* in 1997 if the lamb should be considered seven months old (the period of time since her birth) or six years old (since she was a genetic copy of a six-year-old sheep), Wilmut said, "I can't answer that. We just don't know. There are many things here we will have to find out."

ANIMAL CLONING

Why clone animals in the first place? What are the practical reasons? One is organ farming—the mass production of transgenic animals to harvest their organs and transplant them into humans (*xenotransplantation*). Xenotransplantation experiments in other animal species have not proved successful enough for widespread experimentation with humans, and recent success with reprogramming adult skin cells back to an apparently embryonic state makes researchers hopeful that in the future, organs might be grown using cells from the intended recipient. This would solve the problem of immune rejection, while also preventing animal suffering. (See chapter 5 for an in-depth look at this groundbreaking stem cell research.)

Even if the project of xenotransplantation eventually fizzles, other commercial uses for cloned animals include the following:

1. *Pharming.* Dr. Wilmut's research was funded in part by PPL Therapeutics, a British biotechnology company pursuing the project of *pharming*—the mass production of transgenic animals genetically programmed to produce important biological substances like insulin and blood-clotting factor. If one cow could be genetically programmed to produce a medically useful substance

CC (for Carbon Copy or Copy Cat), the first cloned pet, was born at the Texas A&M College of Veterinary Medicine on December 22, 2001. *(College of Veterinary Medicine, Texas A&M University)*

in her milk, and a herd (or multiple herds) of her clones could be produced, the result would be a massive commercial enterprise to produce these substances.
2. *Pet cloning.* The first pet clone to be produced was the kitten named CC (for Copy Cat or Carbon Copy), produced by researchers at Texas A&M after they failed to clone a dog. Out of 87 attempts with cat embryos, CC was the only one to survive.

    The research was supported by the company Genetic Savings and Clone, a biotech firm that hoped to turn pet cloning into a profitable commercial enterprise when and if cloning technology was ever perfected. The company stored pet DNA, for a fee, to be used at a future date. Critics note that pet cloning is unethical, when millions of abandoned dogs and cats lose their lives in shelters every year, and that it is unrealistic, when developmental factors will undoubtedly result in a different pet than the beloved animal that was cloned.
3. *Commercial agriculture.* Another potential use of animal cloning is the production of herds of agricultural animals with desirable traits (from a commercial perspective). If an especially productive animal (a cow that produces unusually large quantities of milk, for example) could be cloned to produce an entire herd, it could significantly increase profits and limit production costs.

Animal experiments—and their direct implications for the reproductive cloning of humans—are the subject of the next section.

CLONING HUMANS?

"Will There Ever Be Another You?" the headline on the cover of *Time* magazine asked after Dolly's arrival. People expressed resistance—even revulsion—to the idea of human cloning, almost universally. On March 4—not even two weeks after Ian Wilmut's announcement—President Clinton issued an executive order banning the use of federal funds in human cloning research and asking for a moratorium on privately funded efforts. He warned scientists against "trying to play God," asserting that "Each human life is unique, born of a miracle that reaches beyond laboratory science." He believed that "we must respect this profound gift and resist the temptation to replicate ourselves."

Ian Wilmut, when asked by the press if he would ever consider cloning humans, dismissed the idea, telling the *New York Times* that he would "find it repugnant" and *World* magazine (a Christian publication) that "all of us would find that offensive." Dr. Harold Varmus, then director of the National Institutes of Health, told a congressional subcommittee that cloning humans would be "repugnant to the American public." Even so, public fears ran wild that the Dolly experiment was the top of a very slippery slope. "To most people," commented the *New York Times*, "the idea of cloning is frightening; it is evidence of technology speeding out of control, an Orwellian universe where the essence of humanity has been lost and the fact of it has been cheapened."

Not everyone believed that cloning humans would be immoral. University of Tennessee ethicist Carson Strong, for example, argued in his essay "Cloning and Infertility" that if reproductive cloning were ever refined to the degree that it posed no elevated risks to the children created by cloning, it would be an ethically permissible method by which infertile people could have genetically related children. Defenders of human cloning, however, remain a minority voice more than a decade after Dolly's birth. The use of cloning technology on humans remains taboo in mainstream philosophical and medical circles, as evidenced by the unanimous conclusion of President George W. Bush's Council on Bioethics in 2002 that reproductive cloning is unethical. (Council members were split on the issue of the permissibility of therapeutic cloning of human embryos; see chapter 1 for more on the council's decision.)

Some ethicists called for an outright ban on human cloning in the wake of Wilmut's announcement, including the chair of the President's Council on Bioethics from 2002 to 2005, Dr. Leon Kass. In his *New Republic* article, "The Wisdom of Repugnance" (published just a few months after Dolly's existence became public), Dr. Kass wrote that "We are repelled by the prospect of cloning human beings not because of the strangeness or novelty of the undertaking, but because we intuit and feel, immediately and without argument, the violation of things we rightfully hold dear. Repugnance, here as elsewhere, revolts against the excesses of human willfulness, warning us not to transgress what is unspeakably profound." Dr. Kass's intuition is similar to sentiments expressed by opponents of IVF a decade earlier, though in the case of IVF, public aversion to the procedure subsided quickly as it became a common method for overcoming infertility.

Wilmut was taken aback by public fears that his research would lead to the cloning of humans. "People say that cloning means that if a child dies, you can get that child back. It's heart-wrenching. You could never get that child back. It would be something different. You need to understand the biology. People are not genes. They are so much more than that."

The same argument could be made for nonhuman mammals, as CC the cat proved when she was born in 2002. CC, despite her name, is not an exact copy of her genetic mother, Rainbow, in obvious ways; their coats look quite different, since the distribution of pigments responsible for coat colors and patterns occurs during fetal development and is not solely determined by genetic makeup. Even so, the sharp moral divide between cloning animals and cloning humans persists; scientists have succeeded in cloning mice, pigs, cattle, horses, mules, goats, cats, and now even monkeys, encountering very little public resistance along the way.

ECTOGENESIS: GROWING BABIES OUTSIDE THE WOMB

The newest reproductive technology on the horizon is ectogenesis—the gestation of an embryo or fetus in an artificial womb completely apart from a woman's body. If it becomes technically possible, ectogenesis could be used to save extremely premature babies who would otherwise die outside the womb. At its more controversial extreme, the procedure could be used to gestate a baby entirely in vitro—from fertilization with IVF through the full course of embryonic and fetal development.

128   BEGINNING LIFE

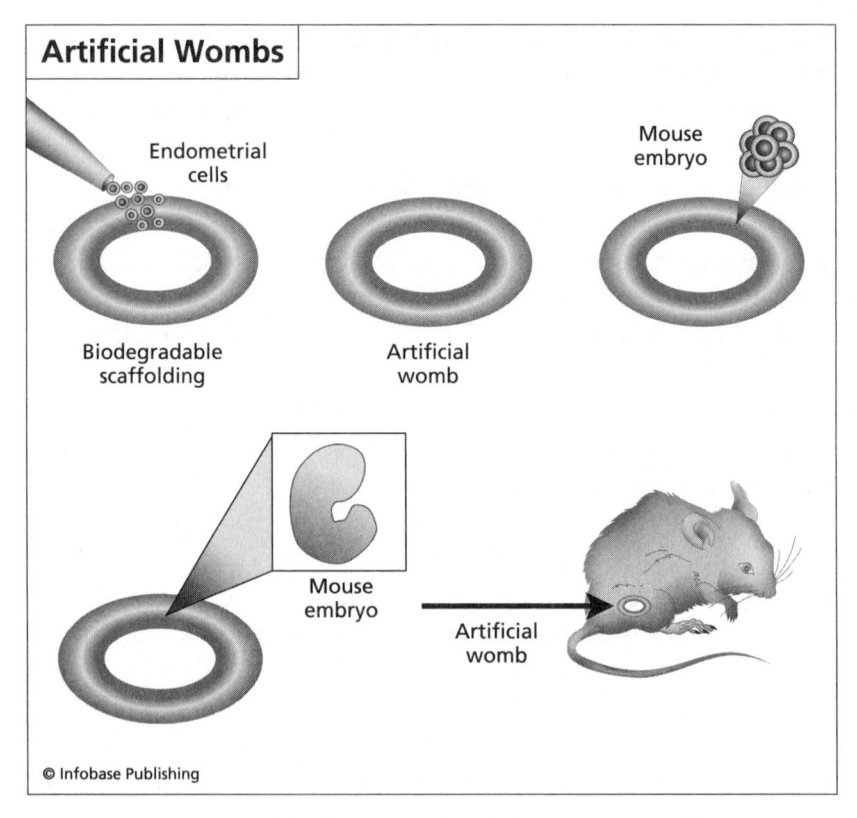

A sketch of Hung-Ching Liu's artificial womb experiments in mice

Ectogenesis may become technically feasible quite soon. Cornell University's Hung-Ching Liu has succeeded in attaching embryos to uterine tissue grown in the laboratory (the embryos were later destroyed). Liu hopes that she will be permitted to extend the length of her experiments to longer gestation periods following more animal tests. Dr. Yoshinori Kuwabara, a Japanese obstetrician, has successfully gestated goats in an artificial womb for three weeks (the equivalent for a goat of one human trimester), while Dr. Thomas Shaffer at Temple University has created an artificial amniotic fluid to help severely premature babies survive. Many premature babies die because their lungs are not developed enough to breathe air, whereas in the womb, "the fetus is in a fluid environment, and its lungs are full of liquid," Dr. Shaffer says. "I am trying to bring the womb environment outside the

ECTOGENESIS AND THE ABORTION DEBATE

Roe v. Wade—the legal pillar for abortion rights in the United States—relies on two basic ideas: first, that every woman has a constitutional right to privacy, and second, that until a fetus has become viable (can survive outside the mother), the woman's right to privacy is paramount. Ectogenesis could—at least in theory—preserve the life of a fetus and a woman's privacy, and in so doing, totally restructure the abortion debate.

The Supreme Court's ruling on *Roe v. Wade* recognized states' rights to regulate abortion, but also specified ways that states could not restrict abortion without violating a woman's constitutional right to privacy. During the first trimester (approximately 12 weeks) of pregnancy, states cannot restrict a woman's right to an abortion at all, and during the second trimester, only restrictions to protect the health and safety of the pregnant woman are allowed. During the third trimester—when the fetus could be considered viable outside the woman's body—restrictions on abortion are permissible, but only if those restrictions preserve a woman's health. (See chapter 6 for more on the case.)

Under current neonatal care conditions, babies born before 25 weeks have less than a 50 percent chance of survival and a significant risk of disability. If ectogenesis became technically feasible, it could make the viability of a fetus outside the woman's body a non-issue, provided that gestation outside a natural womb is not found to be harmful. Ectogenesis could also disable the right-to-privacy case for the second trimester, since any procedure to transfer a fetus to an artificial womb might, arguably, be no more invasive than an abortion. During the first trimester, however, abortion is a relatively minor medical procedure that does not involve surgery or general anesthesia, as opposed to a relatively complicated and invasive "fetal transplant."

(continues)

> *(continued)*
>
> There are still significant technical and ethical barriers to ectogenesis, but observers note that now is the time to consider its ethical and legal ramifications. Attorney Michelle Hibbert argues in her paper, "Artificial Womb Technology and the Constitutional Guarantees of Reproductive Freedom," that although it "may turn out that scientists just won't ever be able to perfect the technology to allow children to be born mechanically," and although "it may turn out that society is unwilling to accept the mechanic bearing of children . . . it is irresponsible to wait until the first child is born of ectogenesis before discussing how the law will, or should, treat that new form of assisted, and collaborative, reproduction."

patient." The results of his experiments, Dr. Shaffer says, indicate that administration of the new fluid to premature infants' lungs could double the survival rate of babies born at 23 weeks (from 35 to 70 percent).

Some ectogenesis advocates also hail the prospect as a blessing for couples who cannot carry their own child—even liberating to women who do not wish to experience pregnancy but who want a child that is genetically their own. Critics warn, however, that a host of legal and ethical problems will emerge as research advances (see the sidebar for possible ramifications for the abortion debate). Even if scientists are able to replicate the highly complex physical environment of a woman's womb—down to healthy levels of oxygen, nutrients, and hormones—there would still be serious ethical barriers to testing the technology for anything other than an emergency intervention, since it will be impossible to know the effects of gestation apart from the mother until after the fact. "We know that a foetus responds to the mother's heartbeat, as well as her emotions, moods and movements," noted Jeremy Rifkin in the British newspaper the *Guardian* in January 2002. "What kind of child will we produce from a liquid medium inside a plastic box? How will gestation in a chamber affect the child's motor functions and emotional and cognitive development? We know that young infants deprived of human touch and bodily contact often are unable to develop the full range of human emotions and sometimes die soon after birth or become violent, sociopathic, or withdrawn later in life."

Despite these and many other concerns, David Magnus, codirector of the Stanford Center for Biomedical Ethics, told CBS Radio's *The Osgood File* in 2004 that he believes that the pressure on scientists to explore the technical limits of ectogenesis will be extreme, and therefore that a regulatory body should be created to deal with ethical issues before they arise. "When things get close [to ectogenesis]," Magnus said, "the temptation to go out and become a cowboy and be the first to be able to do something new will be really strong."

SUMMARY

Beneficence and autonomy are key to any consideration of reproductive cloning and ectogenesis, since the physical and emotional effects on children cannot be established ahead of time. (Unknown risk was a strike against IVF as well, but one that did not prevent the birth of Louise Brown, or of over a million children thereafter.)

Widespread fears that Dolly's birth would inevitably lead to human cloning appear to have been premature, since public revulsion to the idea has so far trumped scientific curiosity. Ectogenesis may receive a warmer welcome if the pressure to save extremely premature newborns is too great to resist. (See the next chapter for issues related to severely impaired newborns.)

Given serious unknown risks to the health and well-being of children produced by new reproductive methods, medical ethicists call for intensive public debates about potential harms and benefits well in advance of human experimentation.

8

Infants

Every year, thousands of new parents are faced with medical decisions on behalf of impaired newborns. The availability of *neonatal intensive care units (NICUs)* since the early 1970s has prolonged—and often saved—the lives of extremely premature or impaired infants. In 2004, the survival rate for babies born after 20 weeks gestation and weighing less than 750 grams was nearly 50 percent, in stark contrast to 20 years ago or more, when a baby weighing less than 750 grams had a slim chance of surviving. Many of these babies grow up to lead healthy, happy lives, while others suffer long-term neurological damage or other serious health problems.

As with so many issues considered in this volume, public policy on the treatment of impaired newborns has developed in a patchwork fashion in reaction to individual cases, leaving most of the difficult decision making up to parents and physicians, and often under extraordinarily taxing circumstances. As a result, some babies have received too little treatment, while others have received excessive treatment given their prognosis. This chapter looks at cases at both extremes, and at efforts since the 1970s to strike a balance between the "undertreatment" and "overtreatment" of impaired newborns.

BABY DOE AND THE TREATMENT OF IMPAIRED INFANTS

In 1982, a child was born with Down syndrome and a condition known as *esophageal atresia* (a closed or underdeveloped esophagus). The baby needed surgery to correct the physical impairment in order to take nourishment. The parents and physicians, however, decided against the surgery, and the baby, who became known as Baby Doe, died six days later.

| **COMMON IMPAIRMENTS SEEN IN NEWBORNS** ||
Type of Impairment	Description and Prognosis
Down syndrome (DS)	A condition typically resulting from an extra chromosome in a person's genome, in which mental retardation and mild physical abnormalities are present. Though children with DS will need special care throughout their lives, they also tend to be happy people. The incidence of Down syndrome is about 1 in 800.
Spina bifida	The most common neural tube defect—affecting about 1,500 babies in the United States every year—spina bifida is a condition in which the fetal spine does not develop correctly, sometimes allowing the spinal membrane and even the spinal cord to protrude. Treatment includes surgery to repair the spine and antibiotics to prevent meningitis (inflammation of the protective membranes around the spine and brain). Some degree of paralysis often results. Spina bifida is frequently accompanied by hydrocephaly.

(continues)

COMMON IMPAIRMENTS SEEN IN NEWBORNS (continued)

Type of Impairment	Description and Prognosis
Anencephaly	A neural tube defect in which the brain of the fetus fails to form properly, resulting in the total absence of the cerebral hemispheres. Anencephalic babies will never be capable of even simple thought, and without aggressive ventilator treatment, they die quickly. Typical treatment includes provision of food and other comfort measures until critical organ systems fail. The rate of occurrence of anencephaly in 2006 was 11.21 per 100,000 live births.
Hydrocephaly	Fluid pressure on the brain, caused by blockage of the flow of cerebrospinal fluid through the spinal canal. Babies can be saved by a surgical procedure to drain the fluid, but brain damage often occurs. This condition frequently results from spina bifida.
Duodenal atresia	A blockage of the upper portion of the small intestine which, in most cases, can be surgically repaired.
Esophageal atresia	A blockage or incomplete formation of the esophagus. Surgery to correct this problem has a high degree of success.

The Baby Doe case turned the undertreatment of impaired infants into a national issue. One month after the baby's death, the Department of Health and Human Services intervened in an unprecedented way: Richard Schultz Schweiker, the secretary of the department, informed hospitals that if they were receiving federal funds, they could not deny "a handicapped infant nutritional sustenance or medical or surgical treatment required to correct a life-threatening condition if (1) the withholding is based on the fact that the infant is handicapped and (2) the handicap does not render treatment or nutritional sustenance contraindicated."

The next four years saw a heated back-and-forth between the executive branch and the federal courts over the appropri-

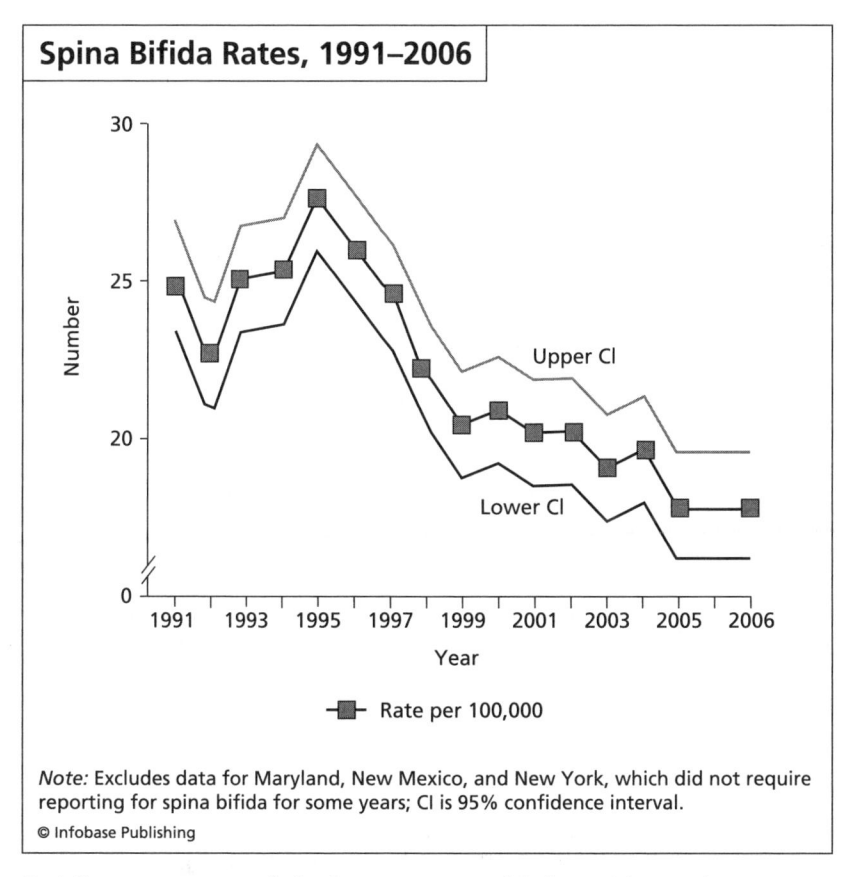

Spina Bifida Rates, 1991–2006

Note: Excludes data for Maryland, New Mexico, and New York, which did not require reporting for spina bifida for some years; CI is 95% confidence interval.
© Infobase Publishing

Public awareness of the importance of folic acid supplementation in preventing some neural tube defects is partly responsible for the decline in rates, as is the wide availability of ultrasound screening, because many pregnancies are terminated out of fear that children with the condition might suffer a poor quality of life. *(Source: CDC, "Trends in Spina Bifida and Anencephalus in the United States, 1991–2006")*

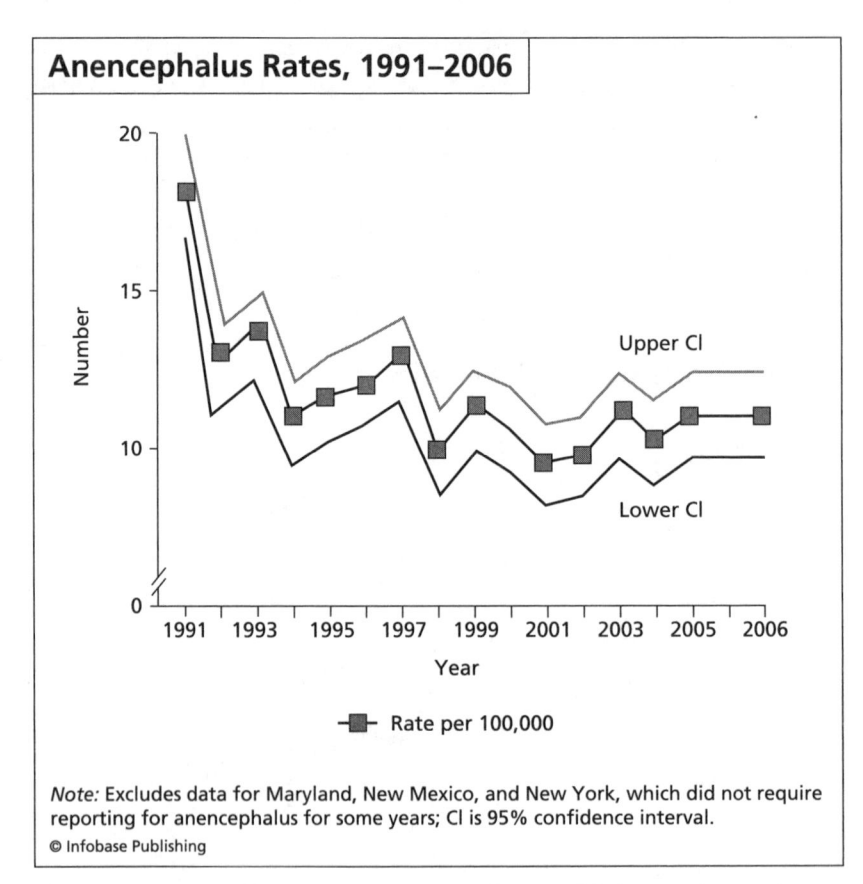

Anencephalus Rates, 1991–2006

Note: Excludes data for Maryland, New Mexico, and New York, which did not require reporting for anencephalus for some years; CI is 95% confidence interval.
© Infobase Publishing

As with spina bifida, the widespread use of folic acid supplements and ultrasound screenings are partly responsible for the decline in new cases. *(Source: CDC, "Trends in Spina Bifida and Anencephalus in the United States, 1991–2006")*

ate role of government in the treatment of impaired infants. Ten months after the HHS secretary's letter, the department—under instructions from President Reagan—issued a detailed set of regulations that required hospitals to display posters notifying the public that discrimination against impaired infants was in violation of federal law, and providing a toll-free, around-the-clock "hotline" number for reporting complaints. Hospitals were required to give HHS officials full access to hospital staff and patient records in response to any alleged violations. A U.S. district court soon invalidated the regulations on the grounds that they should have reflected "caution and sensitivity" in cases involving an impaired child and that "wide public comment prior to rule-making is essential."

After much controversy, public comment, and negotiation among interested groups, the final HHS "Baby Doe" regulations took effect on May 15, 1985. The document defined the term "medical neglect" to include "withholding of medically indicated treatment from a disabled infant with a life-threatening condition." The following year, on June 9, 1986, the U.S. Supreme Court struck down the new regulations, emphasizing in its ruling the lack of evidence that hospitals had been discriminating against handicapped infants, and therefore the lack of grounds for federal intervention. No federal law, the majority noted, required hospitals to treat children in the absence of consent from parents. The Court's ruling essentially placed decision making back in the hands of parents and medical providers, though some states have adopted their own versions of the Baby Doe regulations.

Besides Down syndrome, other common impairments that spark controversy about appropriate treatment include disorders of the spine and brain, such as spina bifida and anencephaly, and digestive tract problems accompanied by mental impairment (see the accompanying table for brief descriptions of these conditions and their prognoses).

EXTREME PREMATURITY

The growing use of fertility drugs and IVF procedures involving multiple embryos has led to an increase in multiple births in this country, and consequently to an increase in the birth of extremely premature, *low birthweight (LBW)* babies. (See chapter 2 for the link between multiple births and fertility treatment.) According to a 2007 report by the CDC, the preterm birth rate rose 2 percent in 2005, accounting for 12.7 percent of all births that year. The percentage of infants born at less than 37 weeks has skyrocketed 20 percent since 1990—and 9 percent since 2000. (Full term is considered to be 40 weeks.)

Premature babies are much more likely to be born at low birth weights, and according to the CDC, the LBW rate rose again to 8.2 percent in 2005—similar to rates seen nearly 40 years earlier. The number of infants weighing less than five pounds eight ounces (2,500 grams) at birth rose 8 percent between 2000 and 2005, and a full 17 percent since 1990. Increases were also recorded for moderately low birth weight babies (1,500–2,499 grams) and very low birth weight babies (less than 1,500 grams, or about three pounds, four ounces).

Very low birth weight babies are at high risk for a variety of complications, including insufficient oxygen levels; breathing problems (such as *respiratory distress syndrome,* a condition caused by immature lungs); neurological problems and brain hemorrhage; inability to regulate and maintain a safe body temperature; gastrointestinal problems (such as *necrotizing enterocolitis,* or *NEC,* a serious disease of the intestine common in premature babies); difficulty feeding and gaining weight; infections;

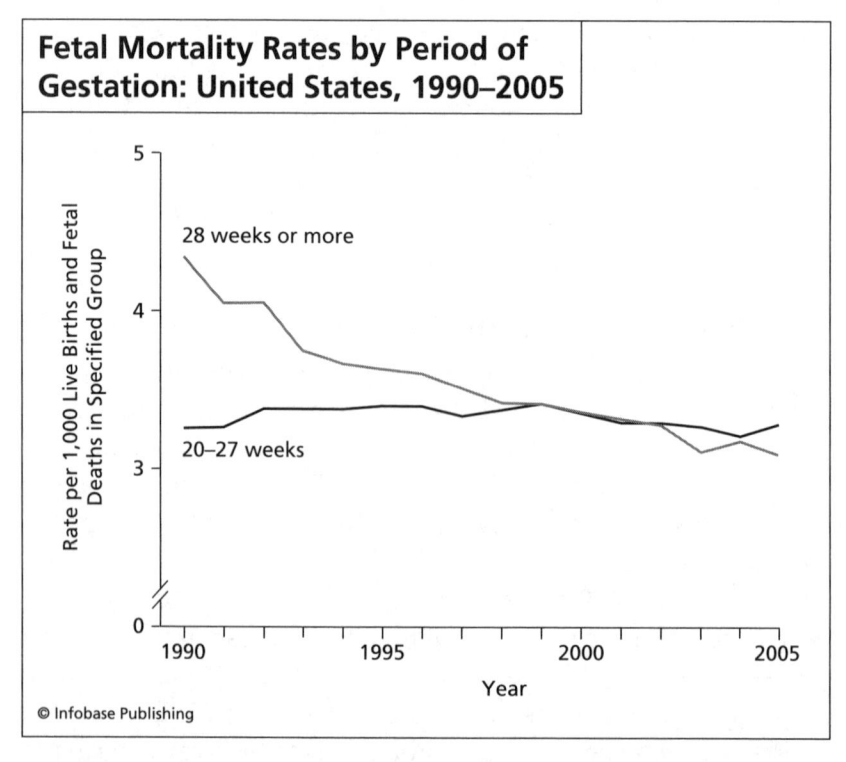

Fetal mortality rates by period of gestation: United States, 1990–2005 *(Source: CDC, National Vital Statistics Reports 57, no. 8, 2009)*

Infants 139

Percentage of Live Births and Infant Deaths by Birthweight in Grams: United States, 2005

Live births

- 2,000–2,499 grams: 5.1%
- 1,500–1,999 grams: 1.6%
- 1,000–1,499 grams: 0.8%
- Less than 500 grams: 0.2%
- 500–999 grams: 0.6%
- 2,500 grams or more: 91.8%

Infant Deaths

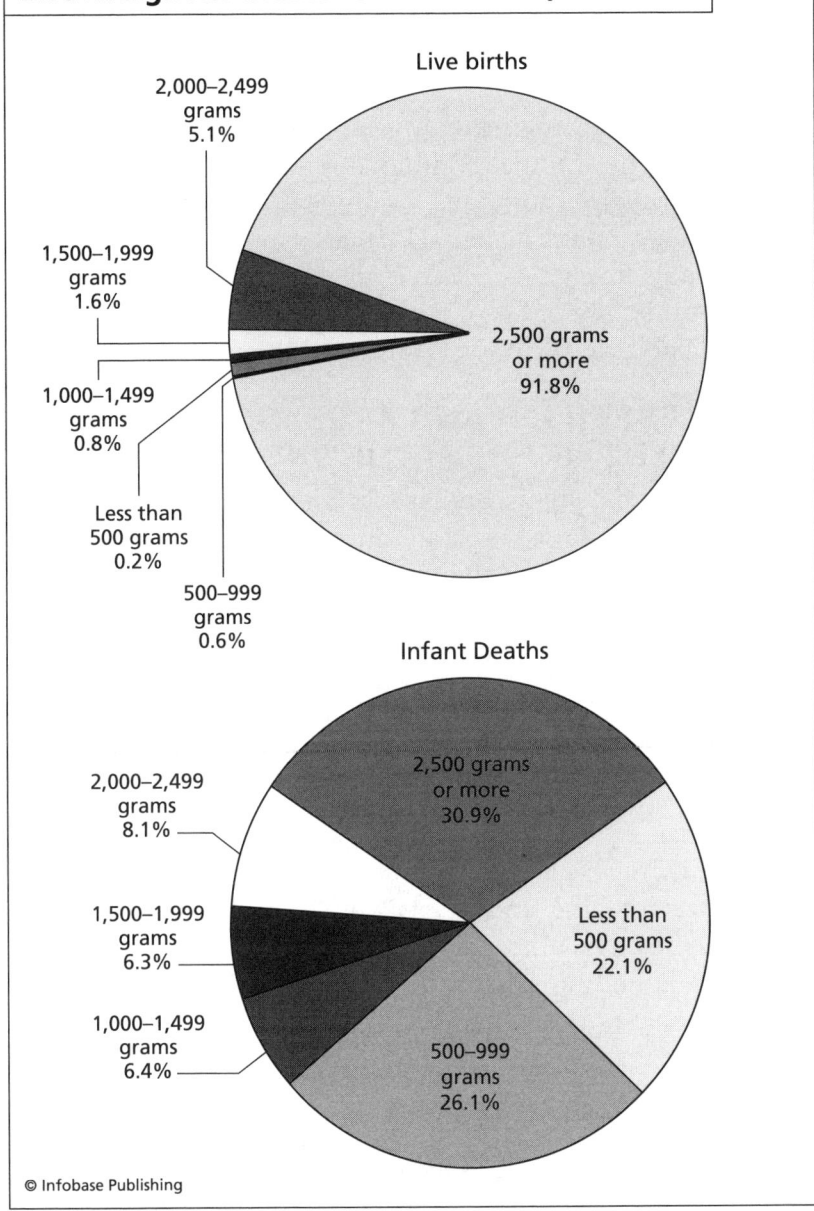

- 2,000–2,499 grams: 8.1%
- 1,500–1,999 grams: 6.3%
- 1,000–1,499 grams: 6.4%
- 500–999 grams: 26.1%
- Less than 500 grams: 22.1%
- 2,500 grams or more: 30.9%

© Infobase Publishing

Percentage of live births and infant deaths by birth weight in grams: United States, 2005 *(Source: CDC, National Vital Statistics Reports 57, no. 8, 2009)*

and *sudden infant death syndrome (SIDS)*. If very low birth weight babies survive, they are at increased risk for long-term problems like cerebral palsy, mental retardation, blindness, and deafness. Generally, the lower the birth weight, the greater the risk for long-term problems.

The Baby Doe regulations grew out of concern that the undertreatment of impaired infants was causing the deaths of babies who might lead happy, fulfilling lives, but the resulting fear of scrutiny and of potential lawsuits caused many physicians and hospitals to aggressively treat newborns who were likely to suffer poor quality of life. The aggressive treatment of severely impaired babies in the NICU is the subject of the next section.

BABY MESSENGER AND AGGRESSIVE TREATMENT OF PREMATURE NEWBORNS

An hour after the premature birth of his son, Dr. Gregory Messenger, a dermatologist on staff at the hospital where the baby was being treated, asked the nurses to leave. He unhooked his son's life support system, and an alarm sounded. The baby was allowed to die in his parents' arms, but the police were notified. Six months later, Dr. Messenger was in a Michigan courtroom standing trial for manslaughter.

Traci Messenger had delivered the baby by emergency cesarean at just 25 weeks gestation. The Messengers were told that the baby had a 50 to 75 percent chance of dying, and that if he survived, there was a 20 to 40 percent chance he would suffer brain hemorrhaging and neurological problems. Respiratory complications were likely as well. Given the odds, the parents asked the attending doctors not to take extraordinary measures to prolong the baby's life. The neonatologist, Dr. Padmoni Karna, gave instructions to intubate the baby (place him on ventilator support) only if he was vigorous and active, and sadly, he was born *hypotonic* (with floppy, loose muscles) and *hypoxic* (lacking sufficient oxygen). He weighed only one pound, 11 ounces. In spite of Dr. Karna's orders, the baby was resuscitated and intubated. Dr. Messenger, heartbroken that his and his wife's wishes for the baby had been ignored, removed the baby from life support.

The county prosecutor, Donald E. Martin, said that although state laws usually allow parents to make medical decisions for their children—including removing them from life support—he decided to prosecute Dr. Messenger because he did not wait for the results of any tests. "The father appeared to make a unilateral decision to end life for his infant son," said Mr. Martin. "Was his act in the best interests of the child? Had he allowed more medical tests, he and his wife would have been in a better position to evaluate the situation. But he took things into his own hands."

After the baby's death, blood-oxygen tests from birth came back at 14 percent—far lower than the 50 percent threshold below which an infant can easily suffer brain damage after five minutes. Dr. Karna testified that given these results, she would have authorized discontinuing life support, but that Dr. Messenger acted on his own before the results were in.

The Messenger case is a clear illustration of the "initiate and reevaluate" approach to treatment—starting aggressive treatment immediately, then reevaluating the course of treatment in light of test results and family wishes—which has become default NICU protocol in the United States and the United Kingdom. "Many bioethicists say they prefer to start treatment, get information and if it looks like conditions warrant stopping treatment you stop," Dr. John Lantos, a pediatrician and associate director of clinical medical ethics at the University of Chicago Medical School, told the *New York Times*. "In actual practice, just the opposite happens. It feels a lot different stopping treatment than it does not to start. If that's what the father feared, that probably reflects a pretty astute understanding of the ethos of most neonatal units." Other countries approach treatment decisions differently; Denmark, for example, uses a statistical system and withholds treatment from newborns who do not meet a threshold of maturity (as detailed in the sidebar).

The controversial gray area between clear harms and benefits, together with the quick pace of medical advancement, have made forging public policy anything but straightforward. There are cases on record in which a hospital advocated withdrawing treatment from a premature newborn, the family insisted on treatment, and the baby turned out to be perfectly healthy. There have also been agonizing cases in which parents advocated withdrawing treatment, hospitals refused, and the baby was left with severely debilitating mental and/or physical conditions. Jan Anderson's son Aaron, for example, was born at 23 weeks gestation and grew to be quadriplegic and virtually blind, with cerebral palsy and perhaps a permanent mental disability. "There is no need for anyone to suffer like this," Aaron's mother told the *New York Times* in September 1991.

State laws enacted after the Baby Doe controversy require the aggressive treatment of older babies and babies born with spina bifida or Down syndrome, but the Messenger case was the first legal test for the treatment of extremely premature infants. Said Arthur L. Caplan, director of the University of

DENMARK'S REQUIRED MINIMUM GESTATIONAL AGE

In stark contrast to the "initiate and reevaluate" approach in the United States and United Kingdom, the Danish Council of Ethics endorses a statistically based approach that accounts for gestational age and physical maturity in decisions about whether to pursue treatment. This protocol dictates that infants born at less than 24 or 25 weeks gestation will not receive aggressive treatment, with the following two qualifications:

1. *Physical maturity.* Under the protocol, even infants of less than 24 or 25 weeks gestational age may be revived, if this is possible using what the council termed "low technology modalities" and minimal handling.
2. *Parental wishes.* Gestational age and maturity criteria may be trumped if parents express willingness to provide necessary care for a premature baby who fails to meet the criteria, or if parents request that

Pennsylvania's bioethics center, "I think what the father did was terribly wrong, understandable in some way, and something that should be looked at very closely by the legal system with a great deal of pity. Given the desperate straights this child was in, it's very hard to find judges and jury who will view the actions taken as more than an emotional response to tragedy or a person pushed to the edge of despair."

Dr. Caplan's prediction was correct: The judge found that there was enough evidence to proceed with a trial, but Dr. Messenger was acquitted of any wrongdoing in February 1995, nearly a year after his son's death.

"Like we said from the very beginning, the hardest part has been losing a child," Dr. Messenger said after the jury's decision. "We're glad to get it over with and move on."

---

doctors withhold treatment from a premature infant who meets the criteria. This qualification is based on the idea that parents must be willing to provide for a newborn's intensive care needs if it is to thrive.

In its 1995 report, "Debate Outline: Extreme Prematurity, Ethical Aspects," the council described two basic principles—the best interests of the baby and economic fairness—on which its recommendations were based: "[T]he panel considers the 35% occurrence of severe handicaps in children born after a pregnancy term of 24–25 full weeks to be high in relation to the number of surviving infants; the panel also takes into account the comparison of the expenditure incurred with the possible alternative applications for that amount." In other words, the high costs of treating extremely premature infants (upwards of $500,000 in the United States for the most premature) could be shifted to the "slightly less premature, since the prospects of better results increase with age and fewer resources are consumed, allowing more to be helped."

Said Mrs. Messenger, "We did what was best for our baby and I will never, ever change my mind."

SUMMARY

The plights of families struggling to care for severely impaired newborns can be heartrending, yet for many premature babies who survive, studies indicate that overall quality of life may be good. A 1996 survey in the *Journal of the American Medical Association* found that 150 premature babies who had reached adolescence—27 percent of whom lived with some disability—rated their quality of life about as high as did a comparison group who were born full term. One teen out of the preterm group reported that death would have been preferable to his present condition.

Treatment decisions on behalf of impaired and extremely premature babies show that the line between benefit and harm shifts as medical knowledge and technology change, and not always in predictable ways. On the one hand, many infants nurtured back from the brink go on to lead happy lives; on the other, some babies who are treated aggressively die after living brief and painful lives. Who should say where to draw the line? This is a tough question always worth asking, and it connects the life-and-death stories in this book.

CHRONOLOGY

ASSISTED REPRODUCTION, STEM CELLS, AND THE ABORTION DEBATE

1869 Pope Pius IX overturns centuries of Catholic dogma, inspired by Thomas Aquinas and, before him, Aristotle, that human "ensoulment" does not occur until the fetus attains an intellectual, rational soul; it is Pius IX who establishes the current official Vatican position that life begins at conception

1951 Pope Pius XII condemns artificial insemination on the grounds that it "converts the domestic hearth, the sanctity of family into nothing more than a biological laboratory"

1962 Phoenix TV personality Sherri Finkbine seeks an abortion when she learns that thalidomide, a tranquilizer she is taking, has been linked to babies being born severely malformed. Ms. Finkbine becomes a lightning rod for angry sentiments from abortion opponents all over the world, and she and her husband eventually fly to Sweden for the procedure

1973 *Roe v. Wade* legalizes abortion in the United States; the Supreme Court's influential ruling recognizes that states can regulate abortion but also specifies ways that states can not restrict abortion without violating a woman's constitutional right to privacy

1974 Federal regulations are promulgated requiring universities or other research institutions that seek or receive federal funds to maintain an Institutional Review Board (IRB), a body charged with reviewing the scientific and ethical merits of federally funded research

1978 The world's first test-tube baby, Louise Brown, is born at Oldham and District General Hospital in north central England, to parents John and Lesley Brown

1979 The National Commission for the Protection of Human Subjects of Biomedical and Behavioral Research releases its seminal Belmont Report

1986 Mary Beth Whitehead gives birth to a baby girl, "Baby M," and refuses to surrender her to William and Elizabeth Stern, the couple with whom she signed a surrogacy contract; Mr. Stern sues to enforce the agreement and to strip Mrs. Whitehead of custody and visitation rights

1987 A New Jersey judge upholds the surrogacy contract and denies Mrs. Whitehead parental rights, despite the fact that she is also the baby's biological mother

1988 The New Jersey Supreme Court overturns the ruling and strikes down the Baby M surrogacy contract, restoring Mrs. Whitehead's parental rights

1990 Anna Johnson, a gestational surrogate, fights to retain custody of the baby she is carrying for Crispina and Mark Calvert; a California Superior Court awards the Calverts full custody and denies Johnson visitation rights on the grounds that she is not the child's biological mother (unlike Mary Beth Whitehead, who was both a genetic and a gestational surrogate in the "Baby M" case)

1997 The McCaughey septuplets are born, setting a world record for the number of babies born alive from one pregnancy

Dolly the sheep's birth is announced; she is the first mammal to be cloned from adult cells

President Clinton issues an executive order banning the use of federal funds in human cloning research and asking for a moratorium on privately funded efforts

1998 Scientists are able to isolate stem cells from human embryos and grow them as cell lines in the laboratory

2001 Helen Beasley, a 26-year-old surrogate mother from England, sues a California couple for allegedly demanding that she abort one fetus when they learn she is carrying twins

2002 India legalizes commercial surrogacy

President George W. Bush's Council on Bioethics unanimously concludes that "cloning-to-produce-children" (also termed reproductive cloning) is unethical; the council remains split on the issue of whether "cloning-for-biomedical-research" (therapeutic cloning) should be allowed

CC, the first cat clone, is born at Texas A&M, to genetic mother Rainbow and surrogate mother Allie

2006 President George W. Bush vetoes the Stem Cell Research Enhancement Act, a bill that would reverse the federal law making it illegal for federal money to be used in research where stem cells are obtained from the destruction of an embryo

The FDA approves over-the-counter sales of the "morning-after pill" for women over 18

Kelly Romenesko, a French teacher from Wisconsin, makes national headlines when she is fired from two Roman Catholic schools for conceiving twins through IVF

2007 The Catholic bishops of England and Wales declare that human-animal embryos conceived in the laboratory for research purposes should be regarded as human life, asserting that if the biological mother wants to carry a hybrid embryo to term, she should be allowed to do so; the bishops are opposed to the creation of any embryo for research but also wish to prevent the destruction of embryos after they have been created

Federal legislation banning intact dilation and extraction (IDX or intact D&X)—or as it is more controversially known in American political debates, partial-birth abortion—is upheld by the Supreme Court in a 5-4 decision

Shinya Yamanaka and colleagues at Kyoto University are able to reprogram mouse skin cells back to an embryonic state, first by introducing 24 genes involved in maintaining plasticity of mouse embryonic stem cells, and then by eliminating those genes one at a time, until four are identified as essential

President George W. Bush vetoes another attempt by Congress to lift restrictions on federally funded stem cell research

Researchers in Britain launch a trial to test stem cell treatment for age-related macular degeneration (AMD); they hope that the "perfect transplant" will be achieved through differentiation of embryonic stem cells into retinal cells

Two teams of scientists—one led by Shinya Yamanaka at Kyoto University, the other by James Thomson at the University of Wisconsin—use the method perfected on mouse skin cells earlier that year to reprogram human skin cells back to an apparently embryonic state

2008 University of Minnesota researchers are able to create a beating rat heart in the laboratory by removing the cells from a dead rat heart, leaving the valves and outer structure intact, and injecting heart cells from newborn rats; the experiments shows that human hearts could one day be grown in the laboratory using stem cells taken from patients' bone marrow

Researchers report success in controlling diabetes in mice by turning human embryonic stem cells into insulin-producing cells that regulate the mice's blood sugar

The British House of Commons defeats a bill that would ban the creation of human-animal hybrid embryos for medical research, thus allowing scientists who obtain proper licenses to create such embryos, harvest their stem cells, and destroy them after 14 days

2009 On the heels of President Barack Obama's inauguration, the FDA approves the world's first human trial with embryonic stem cells (for severe spinal cord injuries)

President Obama issues an executive order lifting the Bush administration's tight restrictions on research with stem cells

Following a ruling by Judge Edward R. Korman of the federal district court in New York, the FDA lowers the age limit of nonprescription sale of the morning-after pill to 17 for both women and men

GENETIC TESTING, GENE THERAPY, AND THE TREATMENT OF IMPAIRED INFANTS

1883 English scientist Francis Galton, cousin of Charles Darwin, coins the term *eugenics*; Galton wishes to improve the human race by eliminating what he calls the "undesirables" and multiplying the "desirables" through the use of arranged marriages

1924 Carrie Buck, 18, is the first person to be chosen for sterilization under a Virginia law requiring sterilization of "mental defectives;" her case prompts a legal challenge that is taken to the U.S. Supreme Court, which ultimately upholds the state's right to forcibly prevent her from conceiving another child

1953 James Watson and Francis Crick first describe the double helix structure of deoxyribonucleic acid (DNA)

1962 Robert Guthrie develops an easy, inexpensive procedure to test infants' blood for the dangerous metabolic disease, PKU, and Massachusetts quickly passes the first mandatory testing law; within four years, PKU testing is required in 41 states

1982 A child known as Baby Doe is born with Down syndrome and a condition known as esophageal atresia (a closed or underdeveloped esophagus); the baby needs surgery to correct the physical impairment in order to be able to take nourishment, but the parents and physicians decide not to pursue surgery and the baby dies six days later

1985 The HHS "Baby Doe" regulations take effect, defining the term "medical neglect" to include "withholding of medically indicated treatment from a disabled infant with a life-threatening condition"

1986 The U.S. Supreme Court strikes down the Baby Doe regulations, emphasizing in its ruling that there is a lack of evidence that hospitals are discriminating against handicapped infants, and therefore that there are no grounds for federal intervention

1988 The U.S. Patent Office grants a patent to the President and Fellows of Harvard College for OncoMouse and other "transgenic nonhuman" mammals

1990 The Human Genome Project—backed by the National Institutes of Health (NIH), the Department of Energy, and other government agencies around the world—undertakes sequencing of the human genome

1994 Dr. Gregory Messenger stands trial for the manslaughter of his premature son after he unhooks his life support system; the child was born 15 weeks early, and his parents were told that he had a 50 to 75 percent chance of dying and that if he survived, there was a 20 to 40 percent chance that he would suffer brain hemorrhaging and neurological problems

1995 Dr. Messenger is acquitted of wrongdoing

The Danish Council of Ethics endorses a statistically based approach whereby infants born at less than 24 or 25 weeks gestation will not receive aggressive treatment unless special conditions are met

China's Maternal and Infant Health Care Law goes into effect—a law designed to actively "prevent new births of inferior quality"

1999 After his participation in a gene-therapy experiment at the University of Pennsylvania, 18-year-old Jesse Gelsinger dies suddenly from what investigators suspect is a severe immunological attack on the experimental viral vector; the FDA puts a temporary halt to two similar gene-therapy trials in humans

Terri Seargent loses her job and health insurance when her employer learns of her potentially expensive medical condition; her case brings national attention to the issue of genetic discrimination

2001 The Indian Supreme Court orders strict adherence to a ban on prenatal gender screening in the hopes of eliminating the practice of sex-selective abortion

2002 Two children in Paris develop a leukemia-like condition after receiving experimental gene therapy for SCID (severe combined immune deficiency, or "bubble boy" disease), while eight of the 11 patients participating in the trial are considered cured; the FDA temporarily halts all similar gene-therapy trials using retroviral vectors in blood stem cells

2003 Sequencing of the human genome is complete

The FDA eases the ban on gene-therapy trials using retroviral vectors in blood stem cells, allowing such trials to proceed for the treatment of life-threatening diseases

2005	The HapMap project gets underway, with the goal of identifying and indexing genetic similarities and differences in humans (known as haplotypes)
2007	The Cancer Genome Atlas (TCGA) project is launched with the mission of identifying genetic variations seen in dozens of types of cancer
	Jolee Mohr, 36, a healthy woman who experienced occasional stiffness from arthritis, dies from a massive fungal infection after participating in an experimental gene-therapy trial for the disease; the tragedy provokes national controversy, although her death has not been linked definitively to the gene-therapy trial
2008	The federal Genetic Information Nondiscrimination Act (GINA) is signed into law

GLOSSARY

abortion spontaneous or intentional termination of a pregnancy before an embryo or fetus can survive independently

adenovirus a type of virus that causes respiratory and eye infections; commonly used to deliver experimental gene therapy

adult stem cell an undifferentiated cell present in adults and children that can renew itself and differentiate (with certain limitations) into the specialized cell types of the tissue in which it was found

age-related macular degeneration a condition that damages the part of the retina responsible for clear central vision, and the leading cause of blindness and impaired vision in people over 60 in the United States and most Western countries

Alzheimer's disease a degenerative brain disease characterized by loss of memory, interference with thinking abilities, and personality changes

AMD see **age-related macular degeneration**

amino acids small organic molecules that are the basic building blocks of proteins

amniocentesis a prenatal test that involves the insertion of a needle through the abdomen to retrieve fluid containing placental cells

amyotrophic lateral sclerosis a debilitating condition in which progressive loss of nerve cells in the spinal cord and brain cause muscle paralysis

anencephaly a neural tube defect in which the brain of the fetus fails to form properly, resulting in the total absence of the cerebral hemispheres

apheresis a procedure in which blood is removed from a patient or donor to separate and extract fluid or cells (such as platelets or stem cells); the blood is then returned to the person's body

Glossary

ART see **assisted reproductive technology**

arthritis inflammation of a joint, often characterized by swelling, pain, and restricted motion

artificial insemination the injection of semen into the uterus (not through sexual intercourse) in order to cause pregnancy

assisted reproductive technology any number of procedures that involve the handling of eggs, sperm, and/or embryos to achieve fertilization without sexual intercourse

autonomy the ability of a person to make independent choices

autosomal dominant disease a condition caused by a single defective gene on a nonsex chromosome that does not require combination with another defective gene in order to be expressed

base pair A pair of complementary nitrogen-rich bases—adenine-thymine or guanine-cytosine—that form each rung on the DNA "ladder"

beneficence one of three central ethical principles articulated in the Belmont Report; in medical ethics, the term refers to the intention on the part of medical providers to act in the best interests of their patients

blastocoel the fluid-filled cavity inside a blastocyst

blastocyst an early-stage embryo of 50–150 cells produced by fertilization and cell division, not yet implanted in the uterine wall

bone marrow the soft tissue at the center of most bones that produces red and white blood cells and platelets

cancer a category of diseases in which abnormal cells divide without control

CD34+ cells blood-forming adult stem cells

cerebral palsy a brain condition that effects muscle coordination and movement

cesarean section a surgical procedure in which a baby is delivered through an incision in the abdomen and uterus

chemotherapy treatment with drugs to kill cancer cells and shrink tumors

chimera an organism composed of two genetically different cell or tissue types, named for the mythological creature that was part lion, part goat, and part serpent

chorionic villus sampling (CVS) procedure for obtaining a small sample of tissue from the placenta (chorionic villi) to perform prenatal genetic tests

chronic myocardial ischemia (CMI) one of the most severe forms of coronary artery disease, in which the tiny vessels that distribute blood throughout the heart muscle become constricted, starving the heart of necessary oxygen

clinical involving the direct observation and treatment of patients

clinical trial an experiment designed to test the effectiveness and safety of a new drug or treatment in humans

cloning-for-biomedical-research see **therapeutic cloning**

cloning-to-produce-children see **reproductive cloning**

COH see **controlled ovarian hyperstimulation**

cones cells of the retina that detect color and fine detail

congenital a condition present at birth; can be due to environmental or genetic factors

control group a group of participants in a clinical trial who receive either the standard treatment for a condition or a "dummy" treatment (placebo) instead of the experimental treatment

controlled ovarian hyperstimulation the use of fertility drugs to stimulate the ovaries to produce multiple eggs in a single cycle

coronary artery disease the most common form of heart disease, in which sticky deposits (plaque) block adequate blood flow to the heart

cytoplasmic hybrid embryo a human embryo created when human DNA is placed into an animal egg from which the nucleus has been removed

deoxyribonucleic acid see **DNA**

depressive disorder a broad term covering many medical conditions characterized by a sense of sadness, hopelessness, or worthlessness, and by a loss of interest in favorite activities; an estimated 20 million American adults live with depressive disorders, and examples include major depressive disorder (clinical depression), postpartum depression, and seasonal affective disorder (SAD)

diabetes a chronic disease in which the body is unable to properly regulate levels of sugar in the blood

differentiation the process by which stem cells give rise to specialized cells of the body

differentiation pathways the developmental routes stem cells follow to produce specialized cell types

directed differentiation differentiation of stem cells into particular cell types

DNA (deoxyribonucleic acid) the chain of molecules inside a cell that carries genetic information and passes it from one generation to the next; it is the chemical blueprint for building, running, and maintaining living organisms

dopamine a neurotransmitter, or chemical produced by the brain that assists in the transmission of electrochemical messages between neurons

Down syndrome a condition typically resulting from an extra chromosome in a person's genome, in which mental retardation and mild physical abnormalities are present

dry macular degeneration the most common type of AMD in which the light-sensitive cells of the macular area begin to deteriorate, resulting in a spotty loss of "straight ahead" vision

duodenal atresia a blockage of the upper portion of the small intestine which, in most cases, can be surgically repaired

ectogenesis gestation of an embryo or fetus in an artificial womb

embolism sudden blockage of an artery by a clot or other material carried by the bloodstream

embryo in humans, the developing organism from the time of fertilization through the eighth week of gestation, at which point it is called a fetus

embryonic stem cell line normal, healthy embryonic stem cells that are proliferating in the laboratory without differentiating into specialized cell types

embryonic stem cells cells from the early embryo with the potential to develop into the specialized cell types of the body

enzyme a protein in the body that enables important chemical reactions

esophageal atresia a blockage or incomplete formation of the esophagus; surgery to correct the problem has a high degree of success

ethical principles general guidelines for ethically appropriate behavior; for example, the Belmont Report outlined three principles by which to evaluate medical research with human subjects (respect for persons, beneficence, and justice)

eugenics any attempt to alter the human species by maximizing the occurrence of certain genetic traits or minimizing the occurrence of others

feeder layer a layer of cells used to maintain stem cells in culture by releasing nutrients and providing a surface to which stem cells can attach

feeding tube a tube used to deliver nutrients to patients who are unable to swallow; examples include the PEG tube, NG tube, and jejunostomy tube

fetus the developing human organism after the eighth week of gestation

full term a pregnancy carried to 40 weeks

futile treatment life-extending treatment that is deemed inappropriate because it will prolong suffering without meaningful hope of recovery, or—in the case of certain profound brain injuries—will sustain certain systems of the body without meaningful hope of regained awareness

GC see **gestational carrier**

gene therapy a collection of techniques designed to correct or replace a gene that is not operating normally

genetic engineering artificially introducing changes to an organism or cell's genetic material

genetic enhancement an addition or alteration to an individual's genome viewed by that individual as a genetic improvement

genetic hybrid see **chimera**

genetic mosaic   see **chimera**

genetic screening   often used interchangeably with **genetic testing**, this term is also used more narrowly to refer to testing populations (or embryos, fetuses, or individuals from those populations) who are at risk for certain diseases

genetic surrogacy   when a surrogate mother also donates the egg that results in pregnancy, making her the biological mother as well as gestational surrogate

genetic testing   testing the DNA of an embryo, fetus, person, or group of people for the presence or absence of particular genes

genome   the complete genetic material of an organism

germ-line therapy   alteration of genes in reproductive cells (sperm or egg) in order to affect their function in any offspring that may be created

gestational carrier   a surrogate mother who carries a pregnancy for the intended parent(s) and who is not the biological mother of the fetus

gestational surrogate   see **gestational carrier**

haemochromatosis   a fairly common and treatable inherited disease characterized by abnormal iron metabolism and caused by mutation of a single gene

haplotype   a group of alleles (alternate forms) of different genes linked closely enough to be inherited as a unit

healthy volunteer   a medical research study participant who is not a patient

heart disease   a general term for a number of conditions that affect the heart—the most common of them coronary artery disease, which is the leading cause of death in the United States

hematopoietic stem cells   see **CD34+ cells**

hemoglobin   the protein in red blood cells that carries oxygen and gives blood its red color

hemophilia   an inherited bleeding disorder caused by a deficiency or abnormality in a protein essential for clotting

hemorrhage   heavy bleeding

hepatitis   inflammation of the liver, caused by infection or toxic agents

homologous recombination a process by which one DNA segment is replaced by another segment with a similar sequence

Huntington's disease a genetic, degenerative disease of the brain and central nervous system that causes progressive loss of mental function and motor control

hydrocephaly fluid pressure on the brain, caused by blockage of the flow of cerebrospinal fluid through the spinal canal

hypothyroidism a condition in which the thyroid gland does not produce enough thyroid hormone, treatable with hormone replacement therapy

hypotonic weak muscle tone

hypoxic lacking sufficient oxygen

IDX see **intact dilation and extraction**

immune response the activity of the immune system against foreign substances or organisms in the body

induced pluripotent stem cells see **iPS cells**

informed consent voluntary consent by a patient or healthy volunteer to participate in a medical research trial, or voluntary consent by a patient to undergo medical treatment, after having been fully informed of—and (ideally) having understood—the anticipated risks and benefits of the treatment or experiment

inner cell mass the cluster of cells inside the blastocyst that is removed to grow embryonic stem cells in culture

insulin a hormone that regulates blood sugar, produced by specialized cells in the pancreas

intact dilation and extraction an abortion procedure in which the brain of a fetus is evacuated and the body delivered vaginally

in vitro fertilization a procedure in which a ripe ovum (egg) is removed from a woman's ovary and placed in a solution of nutrients and sperm to be fertilized; once the ovum has begun to divide into multiple cells, it is implanted in the womb

iPS cells pluripotent stem cells created by genetically reprogramming adult cells back to an embryonic-like state

IRB (institutional review board) a committee formally charged with ethical and scientific review of biomedical and behavioral research involving human subjects; IRB review is required for all human subject research receiving federal funds

IVF see **in vitro fertilization**

justice one of three central ethical principles articulated in the Belmont Report; in medical ethics, the term usually refers to distributive justice—the idea that social benefits and burdens should be fairly shared

LBW see **low birth weight**

Leber's congenital amaurosis a type of inherited blindness that appears at birth or in the first few months of life and prevents the retina from detecting light correctly; it causes progressive deterioration in eyesight and has no known treatment

leukemia cancer of the body's blood-forming tissues, in which the bone marrow produces a large amount of abnormal white blood cells that do not function properly; leukemia means "white blood" in Greek

live birth birth of a living fetus

low birth weight a birth weight of less than five pounds, eight ounces (2,500 grams)

macular area the central part of the retina (the inner lining of back of the eye) rich in cells that detect color and fine detail

metastatic melanoma melanoma (a form of skin cancer) that has spread to other parts of the body

minimal risk when the degree of risk from taking part in a medical experiment is thought to be small—no greater than the risks typically encountered in daily life or in the course of routine medical examinations or tests

moral status the level of consideration given to the interests of a specific entity, e.g. animals or embryos

moratorium a temporary ban or suspension of a specific activity

morning-after pill an emergency contraceptive pill that prevents pregnancy by preventing ovulation (release of an egg from the ovary) or by preventing fertilization of the egg

multiple birth delivery of more than one fetus from a single pregnancy

muscular dystrophy a group of genetic disorders resulting in muscle degeneration and weakness

mutation any change in the DNA of a cell (harmful, beneficial, or with no effect); mutations may be caused by mistakes during cell division or by exposure to DNA-damaging agents in the environment
myelogenous leukemia a disease in which an abnormal protein causes the bone marrow to produce too many white blood cells (all of which contain the mutated chromosome that produce the abnormal protein); eventually the abnormal cells can crowd out healthy blood cells
myeloid blood diseases blood diseases caused by abnormally functioning bone marrow
necrotizing enterocolitis (NEC) a serious disease of the intestine common in premature babies
neonatal refers to the first four weeks of an infant's life
neonatal intensive care unit see **NICU**
NICU a special hospital unit for premature and seriously ill newborns
nontherapeutic experiment an experiment that is not expected to bring therapeutic benefit to participants
oncogene a modified gene capable of transforming normal cells into cancer cells
ornithine transcarbamylase an enzyme critical to the body's ability to rid itself of ammonia
OTC see **ornithine transcarbamylase**
OTC deficiency a metabolic condition in which a lack of the enzyme OTC causes ammonia to accumulate to dangerously high levels in the bloodstream
ovum a female reproductive cell (mature egg) released by the ovary during ovulation
palliative care treatment to improve a patient's quality of life by relieving—rather than attempting to cure—symptoms of a chronic or terminal illness
Parkinson's disease a chronic neurological disorder caused by destruction of dopamine-producing cells in the brain and resulting in loss of motor control
partial-birth abortion see **intact dilation and extraction**
PD see **Parkinson's disease**
perinatal the period just before, during, and after birth

PGD   see **preimplantation genetic diagnosis**

pharming   the farming of genetically engineered animals and plants to produce drugs

phenylketonuria   see **PKU**

PKU   a rare but serious condition that can result in toxic levels of the essential amino acid phenylalanine in the bloodstream, affecting more than one in 25,000 newborns

placebo   an inactive substance or dummy treatment administered to one group of experimental subjects to provide a comparison to the real effects of a test drug or treatment administered to a different group

Plan B   see **morning-after pill**

pluripotent   stem cells that are able to produce all cell types in the body

pre-eclampsia   dangerously high blood pressure induced by pregnancy

preimplantation genetic diagnosis   genetic screening of early embryos for diseases—and sometimes nondisease related traits (like sex)—before implantation in the womb

proliferation   the process by which stem cells are able to renew themselves through cell division for long periods of time

protocol   a written plan for a clinical trial or other experiment that states the purpose of the experiment and exactly how the study will be conducted

regenerative medicine   medical research and treatment focusing on the use of stem cells, altered genes, and growth factors to build new, healthy cells and tissues

reproductive cloning   cloning to produce a child or baby animal

respect for persons   one of the ethical principles articulated in the Belmont Report, which includes respect for the autonomy of people who are capable of making informed, independent decisions and special protections for people whose autonomy is impaired (either due to incompetence or the potential for coercion)

respiratory distress syndrome   a condition occurring primarily in premature infants due to the immaturity of their lungs

retrovirus a type of virus that contains RNA instead of DNA and is able to incorporate itself into the DNA of the host cell

reverse mutation a technique to repair a nonfunctional gene by correcting the specific point where the harmful mutation occurs

RNA (ribonucleic acid) one of the two types of nucleic acids (DNA is the other) found in all cells; RNA contains genetic information necessary to synthesize specific proteins

SCID a rare genetic immune disorder in which the body does not produce healthy versions of the special blood cells that fight infection; also known as "bubble boy" disease

SCNT see **somatic cell nuclear transfer**

seizure a sudden burst of electrical activity in the brain that may produce physical convulsions, thought or sensory disturbances, loss of consciousness, or a combination of symptoms

selective abortion usually refers to the termination of a pregnancy based on the discovery of an undesired, nondisease related trait

selective reduction in a pregnancy with multiple fetuses, a procedure to abort one or more of the fetuses to increase the chances that the others will be born healthy

severe combined immunodeficiency see **SCID**

sex-selective abortion termination of a pregnancy based on the discovery that a fetus is not of the desired sex

sickle-cell disease a group of disorders in which abnormal hemoglobin causes red blood cells to be stiff and abnormally shaped; its effects can vary greatly from person to person and range from almost no ill effects to pain and organ damage, even stroke and death

SIDS the sudden, unexpected death of an apparently healthy baby under one year of age, usually during sleep

single embryo transfer implantation of only one embryo in an IVF procedure to avoid the risk of multiple pregnancy

somatic cell nuclear transfer (SCNT) the process by which a nucleus from an adult (somatic) cell is inserted into an egg whose nucleus has been removed; the first step in reproductive and therapeutic cloning

somatic stem cell see **adult stem cell**

Glossary

spina bifida a condition in which the fetal spine does not develop correctly, sometimes allowing the spinal membrane and even the spinal cord to protrude; the most common neural tube defect, affecting about 1,300 babies in the U.S. every year

spinal tap a procedure in which a fine needle is inserted between two vertebrae in the lower part of the spine to collect cerebrospinal fluid or to administer drugs; also called a lumbar puncture

stem cell an unspecialized cell that can, under the right conditions, develop into the specialized cells of the body

stem cell research research involving embryonic, adult, or induced pluripotent stem cells to study organism development, diseases, and potential treatments

stroke a sudden loss of brain function due to a blockage or rupture in a blood vessel supplying oxygen to the brain

sudden infant death syndrome see SIDS

surrogate motherhood a method of assisted reproduction whereby a woman agrees to become pregnant and deliver a child for another party; she may be the child's genetic mother, or she may be a gestational carrier who is implanted with an embryo fertilized in the laboratory

Tay-Sachs disease a genetic disorder caused by accumulation of a fatty substance in the nerve cells of the brain that results in death in early childhood

telomeres protective regions of DNA at the ends of chromosomes

terminal illness a disease diagnosed as incurable and expected to result in death

terminal sedation (TS) the treatment of pain in terminally ill patients, even to the point of causing unconsciousness or hastening death, usually by means of continuous intravenous administration of a sedative drug; also known as palliative sedation

thalidomide a tranquilizing drug that can cause severe birth defects in a developing fetus when taken by the mother

therapeutic any procedure or drug that is expected to provide medical benefit to patients or alleviate their symptoms

therapeutic cloning   cloning for the purpose of producing embryonic stem cells for medical research or treatment

transhumanism   a movement that advocates the physical and mental enhancement of humans (and of some other animals) by any available technological means

trimester   one of the three approximately three-month periods into which pregnancy is divided

trisomy 18   a rare chromosomal disorder in which an individual has an extra 18th chromosome; the condition is characterized by physical abnormalities, heart problems, and severe mental retardation and is usually fatal within several weeks or months of birth

trophoblast   The outer layer of cells in the blastocyst

ultrasound   a diagnostic imaging technique that uses high-frequency sound waves to generate images of internal body structures

unspecialized   cells that do not have the tissue- or organ-specific structures that allow other cells of the body to perform specialized tasks

ventilator   a machine that assists or maintains breathing for patients unable to breathe on their own; also called an artificial respirator

viral vector   a virus that has been genetically modified in the laboratory to deliver new genes into cells

X chromosome   one of two sex chromosomes (X and Y); human females normally have two X chromosomes and males normally have one X and one Y

xenotransplantation   transplantation of organs or tissues from an organism of one species into an organism of a different species

FURTHER RESOURCES

Chapter 1: Ethical Principles in Genetic and Reproductive Research

Information for this chapter is drawn largely from press reports and journal articles about the birth of Louise Brown and present-day controversies over IVF; historical and policy documents such as the Belmont Report and the President's Council on Bioethics 2002 report on human cloning; and the PBS program, *Test Tube Babies*. All sources are detailed below.

The Associated Press. "Teacher says she was fired over in vitro." September 28, 2006. Available online. URL: http://www.msnbc.msn.com/id/12738144. Accessed October 19, 2009.

BBC News. "Profile: Louise Brown." July 24, 2003. Available online. URL: http://news.bbc.co.uk/1/hi/health/3091241.stm. Accessed October 19, 2009.

Centers for Disease Control and Prevention (CDC). "2006 Assisted Reproductive Technology (ART) Report." November 2008. Available online. URL: http://www.cdc.gov/ART/ART2006/. Accessed October 19, 2009.

Merz, Jon F., Antigone G. Kriss, Debra G. B. Leonard, and Mildred K. Cho. "Diagnostic Testing Fails the Test." *Nature* 415 (February 7, 2002).

Munson, Ronald, ed. "Reproductive Control." Chapter 6 in *Intervention and Reflection: Basic Issues in Medical Ethics*. 7th ed. Belmont, Calif.: Wadsworth/Thomson Learning, 2004.

National Public Radio. "'Left-Over' Embryos Present Dilemma." October 1, 2005. Available online. URL: http://www.npr.org/templates/story/story.php?storyId=4931567. Accessed October 19, 2009.

Newsweek. "All About That Baby." August 7, 1978.

New York Times. "How Are Babies Made? Tale of the Test-Tube Doctors." October 23, 2006. Available online. URL: http://www.nytimes.com/2006/10/23/arts/television/23gate.html?scp=1&sq=%22how+are+babies+made%22&st=nyt. Accessed October 19, 2009.

———. "Patenting Life." February 13, 2007. Available online. URL: http://www.nytimes.com/2007/02/13/opinion/13crichton.html?_r=1&scp=1&sq=%22patenting+life%22&st=nyt&oref=slogin. Accessed October 19, 2009.

———. "As Demand for Donor Eggs Soars, High Prices Stir Ethical Concerns." May 15, 2007. Available online. URL: http://www.nytimes.com/2007/05/15/health/15cons.html?sq=As%20Demand%20for%20Donor%20Eggs%20Soars,%20High%20Prices%20Stir%20Ethical%20Concerns&st=nyt&adxnnl=1&scp=1&adxnnlx=1215619674-hv5dEdCqZpAWLrU55fCiAg. Accessed October 19, 2009.

———. "Your Gamete, Myself." July 15, 2007. Available online. URL: http://www.nytimes.com/2007/07/15/magazine/15egg-t.html. Accessed October 19, 2009.

The President's Council on Bioethics. "Human Cloning and Human Dignity: An Ethical Inquiry." Washington, D.C., July 2002. Available online. URL: http://www.bioethics.gov/reports/cloningreport/index.html. Accessed October 19, 2009.

"A Rush of Test-Tube Babies Ahead?" *U.S. News & World Report.* August 7, 1978.

WGBH/The American Experience, PBS. "Test Tube Babies" Transcripts, interviews, and other materials available online. URL: http://www.pbs.org/wgbh/amex/babies/program/index.html. Accessed October 19, 2009.

Chapter 2: Controversies in Assisted Reproduction

For information on the risks of multiple births to infant and maternal health, Ronald Munson's *Intervention and Reflection: Basic Issues in Medical Ethics,* as well as current CDC reports on assisted reproduction and higher-order multiple births, are key sources for this chapter. Issues arising with surrogacy are detailed in historical and recent press reports, especially those from BBC, NPR, *Newsweek,* and the *New York Times.* All sources are detailed below.

Anderson, Elizabeth. "Is Women's Labor a Commodity?" In *Intervention and Reflection: Basic Issues in Medical Ethics,* 7th ed., edited by Ronald Munson, Belmont, Calif.: Wadsworth/Thomson Learning, 2004.

BBC News. "Couple Denies Surrogate Abortion Claim." August 14, 2001. Available online. URL: http://news.bbc.co.uk/1/hi/health/1489956.stm. Accessed October 19, 2009.

———. "Surrogate Fights to Stop Twins Sale." August 15, 2001. Available online. URL: http://news.bbc.co.uk/2/hi/health/1491926.stm. Accessed October 19, 2009.

Centers for Disease Control and Prevention (CDC). "Higher Order Multiple Births Drop for First Time in a Decade." April 17, 2001. Available online. URL: http://www.cdc.gov/nchs/PRESSROOM/01news/multibir.htm. Accessed October 19, 2009.

———. "2006 Assisted Reproductive Technology (ART) Report." November 2008. Available online. URL: http://www.cdc.gov/ART/ART2006/. Accessed October 19, 2009.

CNN.com. "Surrogate Mother Sues California Couple." August 13, 2001. Available online. URL: http://archives.cnn.com/2001/LAW/08/13/surrogate.dispute/index. Accessed October 19, 2009.

The Independent. "Surrogate Mother Sues over Demand for Abortion." August 11, 2001. Available online. URL: http://www.independent.co.uk/news/world/americas/surrogate-mother-sues-over-demand-for-abortion-665395.html. Accessed October 19, 2009.

Klotzko, Arlene Judith. "Medical Miracle or Medical Mischief? The Saga of the McCaughey Septuplets." In *Intervention and Reflection: Basic Issues in Medical Ethics,* 7th ed., edited by Ronald Munson, Belmont, Calif.: Wadsworth/Thomson Learning, 2004.

March of Dimes. "Multiple Pregnancy and Birth: Considering Fertility Treatments." September 2006. Available online. URL: http://www.marchofdimes.com/files/ACOG_ASRM_MOD_ART_Consumer_FINAL_9-14-06 .pdf. Accessed October 19, 2009.

MSNBC.com. "Do IVF Kids Face More Health Risks?" July 21, 2003. Available online. URL: http://www.msnbc.msn.com/id/3076781/. Accessed October 19, 2009.

National Public Radio. "'Wombs for Rent' Grows in India." December 27, 2007. Available online. URL: http://marketplace.publicradio.org/display/web/2007/12/27/surrogate_mothers/. Accessed October 19, 2009.

New Jersey Monthly. "Now It's Melissa's Time." March 3, 2007. Available online. http://web.archive.org/web/20070526004403/http://www.njmonthly.com/issues/2007/03-Mar/babym.htm. Accessed October 19, 2009.

Newsweek. "The Curious Lives of Surrogates." March 29, 2008. Available online. URL: http://www.newsweek.com/id/129594. Accessed October 19, 2009.

New York Times. "Baby M Case Etches a Study in Contrasts." February 17, 1987. Available online. URL: http://query.nytimes.com/gst/fullpage.html?res=9B0DE7DB1E3AF934A25751C0A961948260. Accessed October 19, 2009.

———. "Father of Baby M Granted Custody; Contract Upheld; Surrogacy is Legal." April 1, 1987. Available online. URL: http://query.nytimes.com/gst/fullpage.html?res=9B0DE5DB1231F932A35757C0A961948260. Accessed October 19, 2009.

———. "Father of Baby M Granted Custody; Contract Upheld; the Absence of Guidelines." April 1, 1987. Available online. URL: http://query.nytimes.com/gst/fullpage.html?res=9B0DEEDF1231F932A35757C0A961948260. Accessed October 19, 2009.

———. "Views on Surrogacy Harden after Baby M Ruling." April 2, 1987. Available online. URL: http://query.nytimes.com/gst/fullpage.html?res=9B0DEED71238F931A35757C0A961948260. Accessed October 19, 2009.

———. "Doctors Are Second-Guessing the Miracle of Multiple Births." June 13, 1999. Available online. URL: http://query.nytimes.com/gst/fullpage.html?res=9B01E0DB1339F930A25755C0A96F958260. Accessed October 19, 2009.

———. "A Surrogate Dries Her Tears." December 11, 2005. Available online. URL: http://www.nytimes.com/2005/12/11/fashion/sundaystyles/11LOVE.html?_r=1&oref=slogin. Accessed October 19, 2009.

———. "Health Guide: Infertility in Women." Available online. URL: http://health.nytimes.com/health/guides/disease/infertility-in-women/overview.html. Accessed October 19, 2009.

———. "India Nurtures Business of Surrogate Motherhood." March 10, 2008. Available online. URL: http://www.nytimes.com/2008/03/10/world/asia/10surrogate.html?_r=1&oref=slogin. Accessed October 19, 2009.

Reuters. "Rent-a-Womb Trend Fuels Debate: American Couples Head to India for Cheaper Fertility Services." February 5, 2007. Available online. URL: http://www.free-articlefinder.com/article.php?id=1474&name=Rent-a-womb+trend+fuels+surrogate+debate. Accessed October 19, 2009.

Science Daily. "IVF Technique Enables Pregnancy without Multiple Births, Study Finds." October 4, 2007. Available online. URL: http://www.sciencedaily.com/releases/2007/10/071001145015.htm. Accessed October 19, 2009.

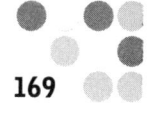

Chapter 3: Eugenics, Genetic Testing, and Designer Babies

The latest on scientific knowledge about the human genome and about genetic links to disease comes primarily from the National Human Genome Research Institute (NIH), and the Francis Collins and Anna Barker article, "Mapping the Cancer Genome." Stephen Jay Gould's essay, "Carrie Buck's Daughter," was a key source for that historical case study, and information on recent controversies and legislation on genetic testing, discrimination, and privacy are drawn largely from current press sources, particularly the *New York Times*. For issues with infant screening for treatable conditions, the March of Dimes Web site is an important research tool. All sources are detailed below.

Brooks, David. "The National Pastime." June 15, 2007. Available online. URL: http://select.nytimes.com/2007/06/15/opinion/15brooks.html?_r=1&oref=slogin. Accessed October 19, 2009.

Collins, Francis S. and Anna D. Barker. "Mapping the Cancer Genome." *Scientific American* 18 (February 2007). Available online. URL: http://www.sciam.com/article.cfm?id=mapping-the-cancer-genome. Accessed October 19, 2009.

Gould, Stephen Jay. "Carrie Buck's Daughter." *The Flamingo's Smile: Reflections in Natural History*. New York: W. W. Norton and Company, 1985.

March of Dimes. "Recommended Newborn Screening Tests: 29 Disorders." April 2006. Available online. URL: http://www.marchofdimes.com/professionals/14332_15455.asp. Accessed October 19, 2009.

Miller, Paul. "Analyzing Genetic Discrimination in the Workplace." *Human Genome News* 12, nos. 1–2 (February 2002). Available online. URL: http://www.ornl.gov/sci/techresources/Human_Genome/publicat/hgn/v12n1/09workplace.shtml. Accessed October 19, 2009.

Munson, Ronald, ed. "Genetic Control." Chapter 5 in *Intervention and Reflection: Basic Issues in Medical Ethics*. 7th ed. Belmont, Calif.: Wadsworth/Thomson Learning, 2004.

National Conference of State Legislatures. "State Genetic Privacy Laws." June 2008. Available online. URL: http://www.ncsl.org/programs/health/genetics/prt.htm. Accessed October 19, 2009.

National Human Genome Research Institute. "Genetic Discrimination." June 19, 2008. Available online. URL: http://www.genome.gov/10002328. Accessed October 19, 2009.

———. "Fact Sheet: Human Genome Project." Available online. URL: http://www.nih.gov/about/researchresultsforthepublic/HumanGenomeProject.pdf. Accessed October 19, 2009.

New York Times. "Senate Sends to House a Bill on Safeguarding Genetic Privacy." October 15, 2003. Available online. URL: http://query.nytimes.com/gst/fullpage.html?res=9C03E3D71E3FF936A25753C1A9659C8B63. Accessed October 19, 2009.

———. "Wanting Babies Like Themselves, Some Parents Choose Genetic Defects." December 5, 2006. Available online. URL: http://www.nytimes.com/2006/12/05/health/05essa.html?adxnnl=1&adxnnlx=1217009 504-xnyQf8S0mKuycleLmbyLjA. Accessed October 19, 2009.

———. "President Calls for Genetic Privacy Bill." January 18, 2007. Available online. URL: http://www.nytimes.com/2007/01/18/washington/18privacy.html?fta=y. Accessed October 19, 2009.

———. "U.S. Set to Begin a Vast Expansion of DNA Sampling." February 5, 2007. Available online. URL: http://www.nytimes.com/2007/02/05/washington/05dna.html?_r=1&scp=1&sq=U.S.%20Set%20to%20Begin%20a%20Vast%20Expansion%20of%20DNA%20Sampling&st=cse&oref=slogin. Accessed October 19, 2009.

———. "Girl or Boy? As Fertility Technology Advances, So Does an Ethical Debate." February 6, 2007. Available online. URL: http://www.nytimes.com/2007/02/06/health/06seco.html. Accessed October 19, 2009.

———. "Genetic Tests Offer Promise, but Raise Questions, Too." February 18, 2007. Available online. URL: http://www.nytimes.com/2007/02/18/business/yourmoney/18reframe.html?ex=1329454800&en=f9496b587cccd74e&ei=5088&partner=rssnyt&emc=rss. Accessed October 19, 2009.

———. "Facing Life with a Lethal Gene." March 18, 2007. Available online. URL: http://www.nytimes.com/2007/03/18/health/18huntington.html. Accessed October 19, 2009.

———. "Insurance Fears Lead Many to Shun DNA Tests." February 24, 2008. Available online. URL: http://www.nytimes.com/2008/02/24/health/24dna.html. Accessed October 19, 2009.

———. "Lawyers Fight DNA Samples gained on Sly." April 3, 2008. Available online. URL: http://www.nytimes.com/2008/04/03/science/03dna.html?scp=6&sq=lawyers%20fight&st=cse. Accessed October 19, 2009.

———. "Congress Near Deal on Genetic Test Bias Bill." April 23, 2008. Available online. URL: http://www.nytimes.com/2008/04/23/business/23gene.html?scp=1&sq=Congress%20Near%20Deal%20on%20Genetic%20Test%20Bias%20Bill&st=cse. Accessed October 19, 2009.

———. "Congress Passes Bill to Bar Bias Based on Genes." May 2, 2008. Available online. URL: http://www.nytimes.com/2008/05/02/health/policy/02gene.html?scp=1&sq=Congress%20Passes%20Bill%20to%20Bar%20Bias%20Based%20on%20Genes&st=cse. Accessed October 19, 2009.

———. "Red Flags for Hereditary Cancers." May 27, 2008. Available online. URL: http://query.nytimes.com/gst/fullpage.html?res=9C01EFDC123DF934A15756C0A96E9C8B63&sec=&spon=&&scp=2&sq=Red%20Flags%20for%20Hereditary%20Cancers&st=cse. Accessed October 19, 2009.

———. "Some Pitfalls of Genetic Testing." May 27, 2008. Available online. URL: http://www.nytimes.com/2008/05/27/health/27brodbox.html?scp=1&sq=Some%20Pitfalls%20of%20Genetic%20Testing&st=cse. Accessed October 19, 2009.

———. "Gene Testing Questioned by Regulators." June 26, 2008. Available online. URL: http://www.nytimes.com/2008/06/26/business/26gene.html?scp=1&sq=Gene%20Testing%20Questioned%20by%20Regulators&st=cse. Accessed October 19, 2009.

Pearson, Veronica. "Population Policy and Eugenics in China." *British Journal of Psychiatry* 167 (July 1995): 1–4.

Scientific American. "Pink Slip in Your Genes." January 18, 2001. Available online. URL: http://www.sciam.com/article.cfm?id=pink-slip-in-your-genes. Accessed October 19, 2009.

Slate. "Better Than Sex: The Growing Practice of Embryo Eugenics." September 16, 2006. Available online. URL: http://www.slate.com/id/2149772/. Accessed October 19, 2009.

———. "The Embryo Factory: The Business Logic of Made-to-Order Babies." January 15, 2007. Available online. URL: http://www.slate.com/id/2157495/. Accessed October 19, 2009.

USA Today. "Newborn Testing Expands to More States, Disorders." July 10, 2007. Available online. URL: http://www.usatoday.com/news/health/2007-07-10-birth-defects_N.htm. Accessed October 19, 2009.

Washington Post. "Increasingly, Couples Use Embryo Screening." September 21, 2006. Available online. URL: http://www.washington

post.com/wp-dyn/content/article/2006/09/20/AR2006092001652.html. Accessed October 19, 2009.

Chapter 4: Gene Therapy and Enhancement

Information on technical aspects of gene therapy comes largely from the U.S. Department of Energy's Human Genome Program and the National Cancer Institute. Jesse Gelsinger's story and accounts of recent gene therapy successes are drawn from a range of relevant press reports. The *Washington Post* is a key source for Jolee Mohr's story. All sources are detailed below.

Cincinnati Children's Hospital. "Gene Therapy Appears to Cure Myeloid Blood Diseases in Groundbreaking International Study." March 31, 2006. Available online. URL: http://www.cincinnatichildrens.org/about/news/release/2006/3-gene-therapy.htm. Accessed October 19, 2009.

National Cancer Institute. "Gene Therapy for Cancer: Questions and Answers." August 31, 2006. Available online. URL: http://www.cancer.gov/cancertopics/factsheet/Therapy/Gene. Accessed October 19, 2009.

———. "New Method of Gene Therapy Alters Immune Cells for Treatment of Advanced Melanoma." August 31, 2006. Available online. URL: http://www.cancer.gov/newscenter/pressreleases/MelanomaGeneTherapy. Accessed October 19, 2009.

National Geographic News. "First Genetically Modified Primate Introduced." January 11, 2001. Available online. URL: http://news.nationalgeographic.com/news/2001/01/0111monkey.html. Accessed October 19, 2009.

New York Times. "Death in Gene Therapy Treatment Is Still Unexplained." September 18, 2007. Available online. URL: http://www.nytimes.com/2007/09/18/health/18gene.html. Accessed October 19, 2009.

Reuters. "Doctors Test Gene Therapy to Treat Blindness." May 1, 2007. Available online. URL: http://www.reuters.com/article/scienceNews/idUSL01665362007050l?pageNumber=1. Accessed October 19, 2009.

Stock, Gregory. *Redesigning Humans: Our Inevitable Genetic Future.* Boston: Houghton Mifflin, 2002.

U.S. Department of Energy. Office of Science. Human Genome Program. "Gene Therapy." June 11, 2009. Available online. URL: http://

www.ornl.gov/sci/techresources/Human_Genome/medicine/
genetherapy.shtml. Accessed October 19, 2009.
Washington Post. "Death Points to Risks in Research." August 6, 2007.
Available online. URL: http://www.washingtonpost.com/wp-dyn/
content/article/2007/08/05/AR2007080501636.html. Accessed
October 19, 2009.
———. "Fungus Infected Woman Who Died After Gene Therapy."
August 17, 2007. Available online. URL: http://www.washington
post.com/wp-dyn/content/article/2007/08/16/AR2007081602322.
html. Accessed October 19, 2009.
———. "Watching Over Clinical Trials." August 17, 2007. Available
online. URL: http://www.washingtonpost.com/wp-dyn/content/
article/2007/08/16/AR2007081602325.html. Accessed October 19,
2009.
———. "Role of Gene Therapy in Death Called Unclear." September 18, 2007. Available online. URL: http://www.washingtonpost.
com/wp-dyn/content/article/2007/09/17/AR2007091701588.html.
Accessed October 19, 2009.

Chapter 5: Stem Cells and Therapeutic Cloning

For scientific and technical basics on stem cells and stem cell research, the NIH's *Stem Cell Information* is an especially rich resource. Information on recent breakthroughs in turning adult cells "embryonic" is drawn largely from recent press reports, especially articles appearing in *Scientific American* and the *New York Times*.

ABC News. "Stem-Cell Research Debate Invades the World Series."
October 26, 2006. Available online. URL: http://abcnews.go.com/
Politics/story?id=2608170&page=1. Accessed October 19, 2009.
The Associated Press. "Celebrities Support Stem-Cell Research."
May 10, 2004. Available online. URL: http://www.signonsan
diego.com/uniontrib/20040510/news_1n10stemcell.html. Accessed
October 19, 2009.
———. "Actors, Athletes to Be in Stem-Cell Ad." October 25, 2006.
Available online. URL: http://www.washingtonpost.com/wp-dyn/
content/article/2006/10/25/AR2006102500145.html. Accessed
October 19, 2009.
CNN.com. "Cells Transformed in Promising Research." August
27, 2008. Available online. URL: http://www.cnn.com/2008/
HEALTH/08/27/cell.identity.ap/. Accessed October 19, 2009.

The Guardian. "Stem Cell Hope for Victims of Age-Related Blindness." June 6, 2007. Available online. URL: http://www.guardian.co.uk/science/2007/jun/06/genetics.medicineandhealth. Accessed October 19, 2009.

National Institutes of Health. Department of Health and Human Services. *Stem Cell Information*. 2009. Available online. URL: Accessed October 19, 2009.

National Public Radio. "States Take Lead in Funding Stem-Cell Research." March 30, 2007. Available online. URL: http://www.npr.org/templates/story/story.php?storyId=9244363. Accessed October 19, 2009.

Nature Reports Stem Cells. "James Thomson: Shifts from Embryonic Stem Cells to Induced Pluripotency." August 14, 2008. Available online. URL: http://www.nature.com/stemcells/2008/0808/080814/full/stemcells.2008.118.html. Accessed October 19, 2009.

New York Times. "Biologists Make Skin Cells Work Like Stem Cells." June 6, 2007. Available online. URL: http://www.nytimes.com/2007/06/07/science/07cell.html?_r=1&oref=slogin. Accessed October 19, 2009.

———. "Bush Vetoes Measure on Stem Cell Research." June 21, 2007. Available online. URL: http://www.nytimes.com/2007/06/21/washington/21stem. Accessed October 19, 2009.

———. "Scientists Bypass Need for Embryo to Get Stem Cells." November 21, 2007. Available online. URL: http://www.nytimes.com/2007/11/21/science/21stem. Accessed October 19, 2009.

———. "Method Equalizes Stem Cell Debate." November 21, 2007. Available online. URL: http://www.nytimes.com/2007/11/21/washington/21bush.html?hp. Accessed October 19, 2009.

———. "Man Who Helped Start Stem Cell War May End It." November 22, 2007. Available online. URL: http://www.nytimes.com/2007/11/22/science/22stem. Accessed October 19, 2009.

———. "Medicine; Suddenly, Stem Cell Central." November 25, 2007. Available online. URL: http://query.nytimes.com/gst/fullpage.html?res=9900E6DF1539F936A15752C1A9619C8B63&scp=1&sq=Medicine;%20Suddenly,%20Stem%20Cell%20Central&st=cse. Accessed October 19, 2009.

———. "After Stem-Cell Breakthrough, the Real Work Begins." November 27, 2007. Available online. URL: http://query.nytimes.com/gst/fullpage.html?res=9E00EFDB173EF934A15752C1A9619C8B63. Accessed October 19, 2009.

———. "Risk Taking Is in His Genes." December 11, 2007. Available online. URL: http://www.nytimes.com/2007/12/11/science/11prof.html?ref=science. Accessed October 19, 2009.

———. "Team Creates Rat Heart Using Cells of Baby Rats." January 14, 2008. Available online. URL: http://www.nytimes.com/2008/01/14/health/14heart. Accessed October 19, 2009.

———. "Using Stem Cells, Researchers Control Diabetes in Mice." February 21, 2008. Available online. URL: http://query.nytimes.com/gst/fullpage.html?res=9506EFD81439F932A15751C0A96E9C8B63. Accessed October 19, 2009.

———. "$271 Million for Research on Stem Cells in California." May 8, 2008. Available online. URL: http://www.nytimes.com/2008/05/08/us/08stem.html?ref=us. Accessed October 19, 2009.

———. "Obama Lifts Bush's Strict Limits on Stem Cell Research." March 9, 2009. Available online. URL: http://www.nytimes.com/2009/03/10/us/politics/10stem.html. Accessed October 19, 2009.

Reuters. "Scientists Plan Stem Cell Cure for Blindness." June 5, 2007. Available online. URL: http://www.reuters.com/article/topNews/idUSL0584115420070605. Accessed October 19, 2009.

Science Daily. "Testing Adult Stem Cells for Heart Damage Repair." March 13, 2007. Available online. URL: http://www.sciencedaily.com/releases/2007/03/070312231717.htm. Accessed October 19, 2009.

Scientific American. "NIH Chief Calls for More Stem Cell Research." March 20, 2007. Available online. URL: http://www.sciam.com/article.cfm?id=nih-chief-calls-for-more&ref=rss. Accessed October 19, 2009.

———. "Dolly's Creator Moves Away from Cloning and Embryonic Stem Cells." July 22, 2008. Available online. URL: http://www.sciam.com/article.cfm?id=no-more-cloning-around. Accessed October 19, 2009.

———. "Is It Time to Give Up on Therapeutic Cloning? A Q&A with Ian Wilmut." July 22, 2008. Available online. URL: http://www.sciam.com/article.cfm?id=therapeutic-cloning-discussion-ian-wilmut. Accessed October 19, 2009.

Spotlight Health. "Dustin Hoffman Advocates Stem Cell Research." May 30, 2004. Available online. URL: http://www.drdonnica.com/celebrities/00008543. Accessed October 19, 2009.

Stateline.org. "Embryonic Stem Cell Research Divides States." June 21, 2007. Available online. URL: http://www.stateline.org/live/details/story?contentId=218416. Accessed October 19, 2009.

University of Texas Health Science Center at Houston. "Stem Cells 101." *Health Leader.* February 7, 2007. Available online. URL: http://publicaffairs.uth.tmc.edu/hleader/archive/RESEARCH/2007/stemcell-0206.html. Accessed October 19, 2009.

University of Wisconsin Hospital and Clinics. "Stem Cells: Healing Hearts." 2007. Available online. URL: http://www.uwhealth.org/news/stemcellshealinghearts/11937. Accessed October 19, 2009.

Chapter 6: Abortion and Emergency Contraception

Data on abortion in the United States are drawn primarily from CDC reports, and information on the right-to-life debate for human-animal hybrid embryos comes primarily from the British press sources, the *Times* and *Telegraph*. The *New York Times* was a key source for recent political debates over emergency contraception and restricted abortion rights. Ronald Munson's chapter on abortion in *Intervention and Reflection*, as well as articles drawn from Helga Kuhse and Peter Singer's text, *Bioethics: An Anthology*, provide a rich spectrum of philosophical viewpoints on abortion and emergency contraception. All sources are detailed below.

Blackstone, William. *Commentaries on the Laws of England: A Facsimile of the First Edition of 1765–1769*. Chicago: University of Chicago Press, 1979. Available online. URL: http://press-pubs.uchicago.edu/founders/documents/amendIXs1. Accessed October 19, 2009.

Bracton, Henry. *On the Laws and Customs of England. Vol. 2*. Cambridge, Mass.: The President and Fellows of Harvard College, 1968–1977. Available online. URL: http://hlsl5.law.harvard.edu/cgi-bin/brac-hilite.cgi?Unframed+English+2+341+abortion. Accessed October 19, 2009.

Butler, Declan. "Conclave Kindles Hope for Bioethical Reform." *Nature* 434 (April 21, 2005): 944.

Centers for Disease Control and Prevention. *Abortion Surveillance—United States, 2005*. November 28, 2008. Available online. URL: http://www.cdc.gov/reproductivehealth/data_stats/Abortion.htm. Accessed October 19, 2009.

Finnis, John. "Abortion and Health Care Ethics." In *Bioethics: An Anthology*, 2nd ed., edited by Helga Kuhse and Peter Singer. Malden, Mass.: Blackwell, 2006.

Telegraph. "Chimera Embryos Have Right to Life, Say Bishops." June 27, 2007.

———. "MPs Vote Against Human-Animal Hybrid Ban." May 19, 2008.

The Times. "Human-Animal Hybrid Embryos Should Be Legal Says Catholic Church." June 27, 2007. Available online. URL: http://

www.timesonline.co.uk/tol/comment/faith/article1993579.ece. Accessed October 19, 2009.

Marquis, Don. "Why Abortion Is Immoral." In *Bioethics: An Anthology*. 2nd ed., edited by Helga Kuhse and Peter Singer. Malden, Mass.: Blackwell, 2006.

Munson, Ronald, ed. "Abortion." Chapter 9 in *Intervention and Reflection: Basic Issues in Medical Ethics*. 7th ed. Belmont, Calif.: Wadsworth/Thomson Learning, 2004.

New York Times. "Debate on Selling Morning-After Pill Over the Counter." December 12, 2003. Available online. URL: http://query.nytimes.com/gst/fullpage.html?res=9E03E5D8143CF931A25751C1A9659C8B63. Accessed October 19, 2009.

———. "Some Doctors Voice Worry over Abortion Pills' Safety." April 1, 2006. Available online. URL: http://www.nytimes.com/2006/04/01/health/01abort.html?_r=1&oref=slogin. Accessed October 19, 2009.

———. "F.D.A. Approves Broader Access to Next-Day Pill." August 25, 2006. Available online. URL: http://www.nytimes.com/2006/08/25/health/25fda.html. Accessed October 19, 2009.

———. "New Push Likely for Restrictions over Abortions." April 20, 2007. Available online. URL: http://www.nytimes.com/2007/04/20/us/20states.html. Accessed October 19, 2009.

———. "Adjudging a Moral Harm to Women from Abortions." April 20, 2007. Available online. URL: http://www.nytimes.com/2007/04/20/us/20assess.html?scp=1&sq=%22reva%20b.%20siegel%22&st=cse. Accessed October 19, 2009.

———. "Abortion Foes See Validation for New Tactic." May 22, 2007. Available online. URL: http://www.nytimes.com/2007/05/22/washington/22abortion.html?scp=1&sq=%22Abortion+foes+see%22&st=nyt. Accessed October 19, 2009.

Sherwin, Susan. "Abortion Through a Feminist Ethic Lens." In *Intervention and Reflection: Basic Issues in Medical Ethics*, 7th ed., edited by Ronald Munson. Belmont, Calif.: Wadsworth/Thomson Learning, 2004.

Chapter 7: Reproductive Cloning and Ectogenesis

Dolly's incredible story comes primarily from press reports at the time of her birth, especially articles appearing in *Time* magazine, *Science* magazine, and the *New York Times*. Perspectives on the emerging science of ectogenesis and artificial womb technology are drawn largely from recent press accounts and from Amel Alghrani's article, "The Legal and Ethical Ramifications of Ectogenesis." All sources are detailed below.

Alghrani, Amel. "The Legal and Ethical Ramifications of Ectogenesis." *Asian Journal of WTO & International Health Law and Policy* 2, no. 1 (March 2007): 189–212.

CBS Radio Network. "Ectogenesis," on *The Osgood File*. June 9, 2004. Available online. URL: http://www.acfnewsource.org/science/ectogenesis.html. Accessed October 19, 2009.

Kass, Leon. "The Wisdom of Repugnance." *The New Republic* 216, no. 22 (June 2, 1997): 17–26. Available online. URL: http://www.catholiceducation.org/articles/medical_ethics/me0006.html. Accessed October 19, 2009.

New York Times. "After Decades of Many Missteps, Cloning Success." March 3, 1997. Available online. URL: http://www.nytimes.com/books/97/12/28/home/0303cloning-sci.html. Accessed October 19, 2009.

———. "Clinton Bans Federal Money for Efforts to Clone Humans." March 5, 1997. Available online. URL: http://query.nytimes.com/gst/fullpage.html?res=9407EFDB1530F936A35750C0A961958260. Accessed October 19, 2009.

Pennisi, Elizabeth, and Nigel Williams. "Will Dolly Send in the Clones?" *Science* 275 (March 7, 1997): 1,415–1,416.

San Francisco Chronicle. "Ectogenesis: Development of Artificial Wombs: Technology's Threat to Abortion Rights." August 24, 2003. Available online. URL: http://www.sfgate.com/cgi-bin/article.cgi?file=/chronicle/archive/2003/08/24/IN273768.DTL. Accessed October 19, 2009.

Strong, Carson. "Cloning and Infertility." *Cambridge Quarterly of Healthcare Ethics* 7 (1998): 279–293.

Time. "Will We Follow the Sheep?" March 10, 1997. Available online. URL: http://www.time.com/time/magazine/article/0,9171,986024,00.html. Accessed October 19, 2009.

World. "Double Double Helix." March 8, 1997.

Chapter 8: Infants

Information on birth defects and on morbidity and mortality for extremely premature newborns comes largely from materials by the National Center for Health Statistics (CDC) and the March of Dimes. For perspectives on the treatment of impaired newborns, Ronald Munson's chapter on impaired infants and medical futility in *Inter-

vention and Reflection: Basic Issues in Medical Ethics is a rich resource. The Gregory Messenger story is detailed in a series of articles from the *New York Times*. All sources are detailed below.

Centers for Disease Control and Prevention (CDC). National Center for Health Statistics. "Trends in Spina Bifida and Anencephalus in the United States, 1991–2006." April 2009. Available online. URL: http://www.cdc.gov/nchs/data/hestat/spine_anen.pdf. Accessed October 19, 2009.

Fost, Norman. "Medical Futility: Commentary." In *Intervention and Reflection: Basic Issues in Medical Ethics,* 7th ed., edited by Ronald Munson. Belmont, Calif.: Wadsworth/Thomson Learning, 2004.

Gross, Michael. "Avoiding Anomalous Newborns." In *Intervention and Reflection: Basic Issues in Medical Ethics,* 7th ed., edited by Ronald Munson. Belmont, Calif.: Wadsworth/Thomson Learning, 2004.

MacDorman, Marian, Martha Munson, and Sharon Kirmeyer. "Fetal and Perinatal Mortality, United States, 2005." *National Vital Statistics Reports* 57, no. 8. Hyattsville, Md.: National Center for Health Statistics, Centers for Disease Control, 2009. Available online. URL: http://www.cdc.gov/nchs/data/nvsr/nvsr57/nvsr57_08.pdf. Accessed October 19, 2009.

March of Dimes. "Birth Defects." April 2009. Available online. URL: http://www.marchofdimes.com/pnhec/4439_1206.asp. Accessed October 19, 2009.

Martin, Joyce, Brady Hamilton, Paul Sutton, Stephanie Ventura, Fay Menacker, Sharon Kirmeyer, and Martha Munson. "Births: Final Data for 2005." *National Vital Statistics Reports* 56, no. 6. Hyattsville, Md.: National Center for Health Statistics, Centers for Disease Control, 2007. Available online. URL: http://www.cdc.gov/nchs/data/nvsr/nvsr56/nvsr56_06.pdf. Accessed October 19, 2009.

Munson, Ronald, ed. "Impaired Infants and Medical Futility." Chapter 10 in *Intervention and Reflection: Basic Issues in Medical Ethics*. 7th ed. Belmont, Calif.: Wadsworth/Thomson Learning, 2004.

New York Times. "Parents of Tiny Infants Find Care Choices Are Not Theirs." September 30, 1991. Available online. URL: http://query.nytimes.com/gst/fullpage.html?res=9D0CE4D71639F933A0575AC0A967958260. Accessed October 19, 2009.

———. "New Medical Quandary at Heart of a Trial." August 3, 1994. Available online. URL: http://query.nytimes.com/gst/fullpage.html?res=9D05E2DE1231F930A3575BC0A962958260. Accessed October 19, 2009.

———. "Father Acquitted in Death of His Premature Baby." February 3, 1995. Available online. URL: http://query.nytimes.com/gst/fullpage.html?res=990CE7DA143FF930A35751C0A963958260. Accessed October 19, 2009.

Web Resources

For the latest information on the reproductive and genetic research and treatment topics considered in this volume, readers can consult the Web sites of the following government agencies, associations, nonprofit organizations, and professional journals. All provide searchable text online and may also provide interactive features, video clips, podcasts, and links to other relevant sources.

AARP (formerly the American Association of Retired Persons). URL: http://www.aarp.org/. Information on a range of topics related to aging. Accessed October 19, 2009.

American Association for the Advancement of Science (AAAS). URL: http://www.aaas.org/. News related to a broad range of scientific topics and careers. Accessed October 19, 2009.

American Psychological Association (APA). URL: http://www.apa.org/. Information on issues and careers in psychology. Accessed October 19, 2009.

American Journal of Bioethics. URL: http://www.bioethics.net/. Free access to abstracts and some full text articles, online discussions, and links to relevant news articles. Accessed October 19, 2009.

Centers for Disease Control and Prevention (CDC). URL: http://www.cdc.gov/. Information on a wide range of health and safety topics. Accessed October 19, 2009.

Hospice Foundation of America (HFA). URL: http://www.hospicefoundation.org/. Information and support resources for people living with terminal diseases and their families and friends. Accessed October 19, 2009.

The Humane Society of the United States (HSUS). URL: http://www.hsus.org/. Information about the treatment and protection of animals in the United States, including animals used in medical research. Accessed October 19, 2009.

Institute for OneWorld Health (iOWH). URL: http://www.oneworldhealth.org/about/index.php. Information on global diseases and the search for cures, from the first nonprofit pharmaceutical company in the United States. Accessed October 19, 2009.

March of Dimes. URL: http://www.marchofdimes.com. Information on prevention of birth defects, premature birth, and infant mortality.

The Mayo Clinic. URL: http://www.mayoclinic.com/. Medical information and online resources from the Mayo Clinic. Accessed October 19, 2009.

National Cancer Institute (NCI). URL: http://www.cancer.gov/. Cancer-related information of all kinds, including descriptions of NCI research programs and clinical trials. Accessed October 19, 2009.

National Institutes of Health (NIH). URL: http://www.nih.gov/. News and information on a wide range of medical research, funding, and career topics. Accessed October 19, 2009.

National Institute of Allergy and Infectious Diseases (NIAID). URL: http://www3.niaid.nih.gov/. Information on a vast array of infectious diseases and their prevention and treatment. Accessed October 19, 2009.

National Institute of Mental Health (NIMH). URL: http://www.nimh.nih.gov/. News and information on a range of issues related to mental health. Accessed October 19, 2009.

Public Library of Science (PLoS). URL: http://www.plos.org/. Open access to articles on a wide range of topics in the medical and biological sciences. Accessed October 19, 2009.

State Children's Health Insurance Program (SCHIP). URL: http://www.cms.hhs.gov/home/schip.asp/. Information on low-cost medical insurance for families and children. Accessed October 19, 2009.

United Network for Organ Sharing (UNOS). URL: http://www.unos.org/. Up-to-date data on organ transplants and available donors. Accessed October 19, 2009.

U.S. Department of Health and Human Services (HHS). URL: http://www.hhs.gov/. Links to information on a variety of health and medical ethics issues. Accessed October 19, 2009.

World Health Organization (WHO). URL: http://www.who.int/en. Information on a wide range of global health topics from the coordinating authority for health within the U.N. system. Accessed October 19, 2009.

World Medical Association (WMA). URL: http://www.wma.net/e/. Information from the global representative body for physicians. Accessed October 19, 2009.

Free Online Print and Radio Media

Several of the print and radio news sources referenced in this volume are available free to online users. The following Web sites contain bonus audiovisual content, such as video clips, slide shows, interactive graphics, and podcasts.

National Public Radio (NPR). URL: http://www.npr.org/. Accessed October 19, 2009.

Newsday. URL: http://www.newsday.com/. Accessed October 19, 2009.

Newsweek. URL: http://www.newsweek.com/. Accessed October 19, 2009.

New York Times. URL: http://www.nytimes.com/. Accessed October 19, 2009.

San Francisco Chronicle. URL: http://www.sfgate.com/. Accessed October 19, 2009.

Slate. URL: http://www.slate.com/. Accessed October 19, 2009.

Time. URL: http://www.time.com/time/. Accessed October 19, 2009.

Washington Post. URL: http://www.washingtonpost.com/. Accessed October 19, 2009.

INDEX

Note: *Italic* page numbers indicate illustrations. Page numbers followed by *t* denote tables, charts, or graphs; page numbers followed by *m* denote maps; page numbers followed by *c* denote chronology entries.

A

AAT (alpha-1 antitrypsin) deficiency 57
abortion 104–112, 115*m*
   and ectogenesis 129–130
   before legalization 105–106
   and McCaughey septuplets 26
   partial-birth 116–118, 147*c*
   and personhood 108–112
   public opinion polls 110*t*
   religious/medical perspectives 106–112
   Roe v. Wade 114–117
   Roman Catholic Church position 106–108, 145*c*
   states likely to protect/restrict, if *Roe v. Wade* is overturned 115*m*
Abse, Leo 61
adenoviruses 65, 68
adult stem cells 88–96, *89*, 90*t*, *91*, *93*, *94*
age-related macular degeneration (AMD) 101–102, 147*c*–148*c*
agriculture, cloning for 125
alpha-1 antitrypsin (AAT) deficiency 57
American Medical Association (AMA) 81
amino acid metabolism disorders 54
amniocentesis 56
amyotrophic lateral sclerosis 96
Anderson, Aaron 142

Anderson, Elizabeth S. 35
Anderson, Jay 142
Anderson, Walter Truett 80
ANDi (rhesus monkey) 77
anencephaly 56, 134*t*, 136*t*
animal cloning *120*, 120–125, *121*, *124*
animal rights 122
Annas, George 25
Aristotle 107, 145*c*
ART. *See* assisted reproductive technology
arthritis 74, 151*c*
artificial insemination 5, 145*c*
artificial lipid spheres 67
assisted reproductive technology (ART) 15*t*, 16*t*, 21–28, *22*. *See also* in vitro fertilization
autosomal dominant disease *59*, 61

B

Baby Doe case 133–137, 149*c*
Baby M case 29–31, *30*, 146*c*
Baby Messenger case 140–144, 150*c*
Barker, Anna D. 44
Barr Pharmaceuticals 114
base pair 43, 63
Beasley, Helen 32–33, 146*c*
Becerra, Xavier 12
Belmont Report 6–7, 146*c*
beneficence 6–7, 25, 29
Benham, Philip 116
Berman, Martha 32, 33

Bernardo, Sanford 33
biogerontologists 46
biotechnology 43–44
Blackstone, William 106–107
blastocoel 85
blastocysts *8,* 85, 98
blood vessels 90
Boersma, Amber 36
bone marrow cells *91,* 148*c*
Botkin, Jeffrey R. 52
Bracton, Henry de 106
breast cancer 60
Brooks, David 53
Brown, Lesley and John *4,* 145*c*
Brown, Louise *xv,* 2–6, *4,* 20, 61, 145*c*
Buck, Carrie 47–49, *48,* 149*c*
Buck, Doris 49
Buck, Emma *48*
Buck v. Bell 47–49, 149*c*
Burger, Warren 10
Bush, George W.
   human cloning 8–9, 126, 146*c*
   human enhancement 79
   IVF 17
   stem cell research 13, 95, 97, 100, 147*c*

C

Caenorhabditis elegans 45–46
California Institute for Regenerative Medicine 100
Calvert, Crispina and Mark 34, 146*c*
cancer 44
cancer stem cells 85
Cantor, Jennifer 36
Caplan, Arthur 6, 16–17, 20, 68, 142–143
Carson, Johnny 13
cat, cloned *124,* 125, 127
Caviezel, Jim 99
CC (cloned cat) *124,* 125, 127
CD34+ cells (hematopoietic stem cells) 88, 91, 92
cerebral palsy 23
cesarean section 22

CF (cystic fibrosis) 56
chimera (genetic mosaic/hybrid) 107–109, 147*c*
China, People's Republic of 50, 52, 150*c*
chorionic villus sampling (CVS) 56
chromosomes *43*
chronic myocardial ischemia (CMI) 90
Claiborne, Ruth 33
Clinton, Bill 125–126, 146*c*
cloning
   for medical research. *See* therapeutic cloning
   to produce children. *See* reproductive cloning
CMI (chronic myocardial ischemia) 90
Coffee, Linda 114
COH (controlled ovarian hyperstimulation) 23
Collins, Francis 58
commercial surrogacy 35–36, 146*c*
congenital defects/disorders 19, 54–56, 133*t*–135*t*
Connell, Noreen 31
Conradi, Wilco 70
contraception, emergency 112–114
controlled ovarian hyperstimulation (COH) 23
coronary artery disease 88, 90
Covington, Sharon 26
Crichton, Michael 11
Crick, Francis 40, *41,* 44, 149*c*
CVS (chorionic villus sampling) 56
cystic fibrosis (CF) 56
cytoplasmic hybrid embryo 108–109

D

da Cruz, Lyndon 101–102
Da Dores, Luciene *105*
Da Dores, Rafael *105*
Davis, Lawrence 78
Denmark, required minimum gestational age in 142–143, 150*c*
differentiation 83, *86,* 87

Index

differentiation pathways 88
directed differentiation 84
DNA (deoxyribonucleic acid)
    animal cloning 122, 123
    gene therapy 66–67, 77
    and genetics 40–43, *41, 42,* 149*c*
Dolly the sheep *120,* 120–123, *121,* 126, 146*c*
donated eggs 16, 18–19
donated embryos 16–18
dopamine 84, 88
double helix 40, 149*c*
Down syndrome 56, 133, 133*t,* 149*c*
Drake, Karen xiv
dry macular degeneration 101
duodenal atresia 134*t*
Dyck, Arthur 5

E

early-stage embryos 9
ectogenesis 127–131, *128*
Edwards, Robert 3, 81
eggs, sale of 18–19
embolism 24
embryonic stem cells 85–88, *86, 87,* 97, 147*c,* 148*c*
embryos. *See also* prenatal screening
    frozen 17–18, *18*
    moral status of 7–9, 12–13
    screening during IVF 50–53, *51*
    and stem cell research 85
    transfer of 27*t*
emergency contraception 112–114
ensoulment 107–108, 145*c*
esophageal atresia 133, 134*t,* 149*c*
ethical principles 1–20
eugenics *46,* 46–57, 148*c*–149*c*
Evans, Roger 117–118
extreme prematurity 137–140

F

fatty acid oxidation disorders 54–55
FDA. *See* Food and Drug Administration
feeder layer 85
feeding tubes 23
Feldt, Gloria 113
feminism 111–112
fertility drugs 22, 24
Finkbine, Sherri 105–106, 145*c*
Finnis, John 107
follow-on studies 75
Food and Drug Administration (FDA)
    embryonic stem cell human trials 100, 148*c*
    emergency contraception 112–114, 147*c*
    gene therapy trials 69, 71, 75, 76, 150*c*
forced sterilization 47
Ford, Harrison 98
Forsythe, Clarke D. 117
Fox, Michael J. 98, 99
France 37
Franz, Wanda 118
frozen embryos 17–18, *18*
Fuller, Barbara 57

G

Galton, Francis 47, 148*c*–149*c*
GC (gestational carrier) 31–33
Gelsinger, Jesse 66–69, 150*c*
Gelsinger, Paul 68, 69
gender selection. *See* sex selection
genes, patenting of 10–12
gene therapy 63–81
    delivery systems 64–67, *65*
    Jesse Gelsinger case 67–69, 150*c*
    genetic enhancement 78–81
    germ-line therapy 76–78
    Jolee Mohr case 72–76, 151*c*
    status of research 69–71
    technical limitations and ethical concerns 71–81
genetic discrimination 57–58, 150*c*
genetic engineering 5–6, 77
genetic enhancement 78–81
Genetic Information Nondiscrimination Act (GINA) 58, 151*c*

genetic mosaic/hybrid 107–109, 147c
genetics 40–46
Genetic Savings and Clone 125
genetic surrogacy 29–31
genetic testing 7. *See also* genetic discrimination
Genomic Research and Accessibility Act 12
Genovo 69
George, Robert 20
Germany 47
germ-line therapy 76–78
gestational carrier (GC) 31–33
gestational surrogacy 31–34, 146c
GINA. *See* Genetic Information Nondiscrimination Act
Ginsburg, Ruth Bader 117
Giuliani, Rudolph 118
Gonzales v. Carhart 117–118
Gould, Stephen Jay 49
Greely, Hank 75
Grifo, Jamie 52
Guthrie, Robert 149c

H
haemochromatosis 11–12
Hanafin, Hilary 39
Handel, William 31
haplotype 44, 151c
HapMap 44, 151c
Harvard College 10, 149c
Hauser, Paula xiv
Hayden, Michael 61
Health and Human Services, U.S. Department of (HHS) 134–137, 149c
health insurance 150c
Heaton, Patricia 99
hematopoietic stem cells. *See* CD34+ cells
hemoglobin 55
hemoglobinopathies 55
hemorrhage 24
Henig, Robin Marantz 13
hepatitis 68
Hereditary Genius (Galton) 47

HHS. *See* Health and Human Services, U.S. Department of
Hibbert, Michelle 130
high-order multiple births xiv, 23–28, 24t
Hitler, Adolf 47
Hoffman, Dustin 98
Holmes, Oliver Wendell 48–49
homologous recombination 64
human-animal hybrids 108–109, 147c, 148c
human cloning 125–127, 146c
human gene pool 76–77
Human Genome Project 43–44, 149c, 150c
human trials 67–69, 72–76, 148c, 150c, 151c
Huntington's disease 58, *60*, 60–61
hybrids, human-animal 108–109, 147c, 148c
hydrocephaly 134t
hypothyroidism 56
hypotonic 140
hypoxic 140

I
ICSI (intracytoplasmic sperm injection) *14*
IDX. *See* intact dilation and extraction
India 36–38, 52, 146c, 150c
induced pluripotent stem cells (iPS cells) 92–97, *93, 94*
infants 132–144
    Baby Doe case 133–137
    Baby Messenger case 140–144
    common impairments seen in 133t–134t
    premature. *See* premature infants
    screening for treatable conditions 54–56
informed consent 19, 25, 72–76
initiate and reevaluate 140
inner cell mass 85, 87
Institute for Human Gene Therapy 67, 69

institutional review board (IRB) 145c
insulin 88
insurance 150c
intact dilation and extraction (IDX) 116–118, 147c
intracytoplasmic sperm injection (ICSI) 14
in vitro fertilization (IVF) 2, 13–20, 14
   Louise Brown 3
   Catholic Church and 5, 16, 147c
   multiple births 26–28
   safety 19–20
   screening 50–53, 51
   selective reduction 32
   for stem cells 85
iPS cells. *See* induced pluripotent stem cells
IRB (institutional review board) 145c
IVF. *See* in vitro fertilization

J
jaundice 68
John Paul II (pope) 98
Johnson, Anna 34, 146c
Johnson, Douglas 102
Johnston, Josephine 19
Jornlin, Marianne 28
justice 7, 29

K
Karna, Padmoni 140
Kass, Leon 126–127
Kennedy, Anthony M. 117, 118
Kennedy, Edward M. 102
Klein, Nathan 66
Klotzko, Arlene Judith 26
Koplan, Jeffrey 25
Korman, Edward R. 148c
Kuwabara, Yoshinori 128

L
Lantos, John 141
Laughlin, Harry 48
LBW. *See* low birth weight

Leber's congenital amaurosis 71
Leigh, Edward 109
leukemia 150c
Levine, Tsila 45
life, patenting of 10–12
Liu, Hung-Ching 128
live birth 14
Lombardo, Paul A. 49
low birth weight (LBW) 19, 137. *See also* extreme prematurity

M
macular area 101
Magnus, David 131
manslaughter 140
Marquis, Don 112
Martin, Donald E. 140
Maternal and Infant Health Care Law (China) 50, 150c
McCaskill, Claire 99
McCaughey, Bobbi xiv, 22–23, 26
McCaughey, Kenny 22, 26
McCaughey septuplets xiv, 21–24, 22, 26, 146c
McCorvey, Norma 114, 116
mechanical ventilators xiv, 23
medical neglect 137, 149c
Mein Kampf (Hitler) 47
Meisner, Andy 100
Mercator Genetics 11–12
Merz, Jon 11, 12
Messenger, Gregory 140, 143, 150c
Messenger, Traci 140, 144
Messenger case 140–144, 150c
metastatic melanoma 71
minimum gestational age 142–143
miscarriage 25
Modafinil 79
Mohr, Jolee 72–76, 151c
Mohr, Rob 73, 74
molecular vectors 67
moral status, of embryos/fetuses 8
moratorium 12
Moreno, Jonathan 75
morning-after pill 112–114, 148c. *See also* Plan B

Moser, Katharine 58, 60, *60*
multiple births xiv, 14, 19, 21–28, 24*t*, 32–33
Munson, Ronald 75, 78
mutation 44
Myers, Gernisha 31–32, 36
Myrah, Steve 90–91

N
National Human Genome Research Institute (NHGRI) 57–58
Nazi Germany 47
necrotizing enterocolitis (NEC) 138
Nelson, K. Ray 49
neonatal intensive care unit (NICU) 132, 141
neurons 88
newborns. *See* infants
NHGRI (National Human Genome Research Institute) 57–58
NICU. *See* neonatal intensive care unit
Niederhuber, John E. 71

O
Obama, Barack 100–102, 148*c*
Omessi, Margalit *45*
oncogene 10
OncoMouse 10, 149*c*
Operation Rescue 116
organic acid metabolism disorders 54
ornithine transcarbamylase (OTC) deficiency 67, 68
ovulation 113
ovum 3

P
Parens, Eric 78, 80
Parkinson's disease (PD) 87–88, 98
partial-birth abortion 116–118, 147*c*
Patel, Nandani 38
patent issues 10–12, 149*c*
PD. *See* Parkinson's disease
pet cloning 125
PGD. *See* preimplantation genetic diagnosis
pharming 124–125
Pius IX (pope) 107, 145*c*
Pius XII (pope) 5, 145*c*
PKU (phenylketonuria) 54, 149*c*
placebo 92
Plan B 112–114
Planned Parenthood 117
pluripotent stem cells 85
PPL Therapeutics 124
pre-eclampsia 24
preimplantation genetic diagnosis (PGD) 28, 50–53, *51*
premature infants
    Baby Messenger 140–144
    Danish Council of Ethics approach 142–143, 150*c*
    extreme prematurity 137–140
    from IVF 19
    live births/deaths by birthweight 139*t*
    McCaughey septuplets xiv
    mortality rates 138*t*
prenatal screening 52, 53, 56–57, 150*c*
President's Council on Bioethics 8, 126, 146*c*–147*c*
privacy 57–58, 129, 145*c*

R
Raper, Steven 68
Raval, Amish 92
Reagan, Nancy 98
Reagan, Ronald 136
Reeve, Christopher 98
regenerative medicine 83
reproductive cloning 119–127
    animal cloning 120–125, *124*
    Dolly the sheep *120*, 120–123, *121*
    human cloning 8–9, 125–127, 146*c*–147*c*
respect for persons 6, 25, 73
respiratory distress syndrome 138
retrovirus 70, 71

Index

Rifkin, Jeremy 130–131
right to life 108–109
right to privacy 129, 145*c*
Robbins, Gregory 107
Roe v. Wade 105, 114–117, 129, 145*c*
Roman Catholic Church
    abortion 106–108, 145*c*
        human-animal hybrids 108–109, 147*c*
    IVF stance 5, 16, 147*c*
Romenesko, Kelly 16, 147*c*
Rosenwaks, Zev 25
Roslin Institute 120, 122, 123
roundworm 45–46
Rupak, Rudy 37–38
Ruse, Cathy Cleaver 113
Ryan, Jennalee 53

S

Saletan, William 53
Sanghavi, Darshak M. 51
Sauer, Mark 26–27
Schweiker, Richard Schultz 134–135
SCID (severe combined immunodeficiency) 70, 150*c*
SCNT. *See* somatic cell nuclear transfer
Scott, Stephanie 33–34
screening
    of donors for IVF 53
    of embryos during IVF 50–53, 51
    for incurable diseases 58–61
    of infants for treatable conditions 54–56
    prenatal 53, 56–57
Seargent, Terri 57, 150*c*
selective abortion 52, 146*c*, 150*c*
selective reduction 23, 32–33
septuplets xiv
severe combined immunodeficiency. *See* SCID
sex selection, PGD for 51–52
sex-selective abortion 52
Shaffer, Thomas 128, 130

Shamoo, Adil 76
Sharma, Priyanka 38
Sherwin, Susan 111–112
Shorett, Peter 11
sickle-cell disease 55
SIDS (sudden infant death syndrome) 140
Siegel, Reva B. 118
single embryo transfer 28
Smith, Father William 4–5
somatic cell nuclear transfer (SCNT) 94, 95
somatic stem cells. *See* adult stem cells
Sorkow, Harvey R. 30, 31
South Dakota 118
Specter, Arlen 95
spina bifida 133*t*, 135*t*
Steinberg, Jeffrey M. 52, 53
Stem Cell Research and Cures Amendment 99
Stem Cell Research Enhancement Act 13, 147*c*
stem cells 82–103, *84*
    adult 88–92, *89*, 90*t*, *91*
    embryonic 85–88, *86*, *87*
    induced pluripotent 92–96, *93*, *94*
    research issues 13, 96–102, 146*c*–148*c*
Steptoe, Patrick 2, 3, 81
Stern, Elizabeth 29, 31, 146*c*
Stern, William 29, 30, *30*, 146*c*
Strong, Carson 126
Studer, Lorenz 93
sudden infant death syndrome (SIDS) 140
Suppan, Jeff 99
Supreme Court, U.S.
    Baby Doe regulations 149*c*
    Buck v. Bell 47–49, 149*c*
    handicapped infant ruling 137
    partial-birth abortion 116, 117, 147*c*
    patenting of life 10
    Roe v. Wade 114–115, 129, 145*c*

surrogate motherhood 28–39, *35*
   genetic surrogacy 29–31
   gestational surrogacy 31–34, 146*c*
   international surrogacy 36–39, *37*
   motives for surrogacy 34–36
   selective reduction 32–33
Switzer, Lisa 38

T

Targeted Genetics 76
Tay-Sachs disease 50, 56
telomeres 123
test-tube babies *xv*, 2–6, 145*c*
thalidomide 105, *105*, 145*c*
The Cancer Genome Atlas (TCGA) 44, 151*c*
therapeutic cloning 9, 147*c*
Thomas Aquinas, Saint 107, 145*c*
Thomson, James 95, 97, 148*c*
transhumanism 79–80
trophoblast 85
Trussell, James 113

U

ultrasound 22, 26
unspecialized cells 83

V

Varmus, Harold 126
Vatican. *See* Roman Catholic Church
ventilators. *See* mechanical ventilators
viral vectors 65–66, 74, 95

W

Wade, Henry 114
Warner, Judith 38
Warner, Kurt 99
Watson, James *41*, 149*c*
Weddington, Sarah 114
Weissman, Irving 96
Weldon, Dave 12
Werb, Zena 122
Wertz, Dorothy C. 57
Wetsel, Rick 100
Wexler, Nancy 61
Wheeler, Charles 32, 33
Whitehead, Mary Beth 29–31, 146*c*
Williams, David 71, 72
Wilmut, Ian 96, 97, 120, 122–124, 126, 127
Wilson, James 69
World Transhumanist Association 79

X

X chromosome 70
xenotransplantation 123

Y

Yamanaka, Shinya 95, 96, 147*c*, 148*c*

Z

Zaner, Richard 25
Zsigmond, Eva 100